T0180130

Synthesis Lectures on Engineering, Science, and Technology

The focus of this series is general topics, and applications about, and for, engineers and scientists on a wide array of applications, methods and advances. Most titles cover subjects such as professional development, education, and study skills, as well as basic introductory undergraduate material and other topics appropriate for a broader and less technical audience.

Gennaro S. Rodrigues ·
Fernanda L. Kastensmidt · Alberto Bosio

Approximate Computing and its Impact on Accuracy, Reliability and Fault-Tolerance

Gennaro S. Rodrigues
Federal University of Rio Grande do Sul
Porto Alegre, Rio Grande do Sul, Brazil

Fernanda L. Kastensmidt
Federal University of Rio Grande do Sul
Porto Alegre, Rio Grande do Sul, Brazil

Alberto Bosio
University Lyon, ECL, INSA Lyon, CNRS,
UCBL, CPE Lyon, INL
Lyon, France

ISSN 2690-0300 ISSN 2690-0327 (electronic)
Synthesis Lectures on Engineering, Science, and Technology
ISBN 978-3-031-15719-6 ISBN 978-3-031-15717-2 (eBook)
https://doi.org/10.1007/978-3-031-15717-2

This Springer imprint is published by the registered company Springer Nature Switzerland AG
The registered company address is: Gewerbestrasse 11, 6330 Cham, Switzerland

The truth is not for all men but only for those who seek it.

—Ayn Rand

Preface

This book presents a study on the reliability of embedded systems when approximate computing is applied to both the software and hardware layers. It aims to introduce fault tolerance techniques based on approximate computing and proves that approximate computing can be applied to safety-critical applications. More in detail, the book overviews approximate computing methods and proposes approximate fault tolerance techniques applied to reconfigurable hardware and embedded software to ensure high reliability level at low computational costs.

Inspired by the trade-off between the improvements provided by approximate computing and the safety-critical system requirements, the book presents an analysis of the applicability of approximate computing techniques to safety-critical systems. The proposed techniques are tested under simulation, emulation, and hardware-based fault injection experiments. Results show that approximate computing algorithms have a particular behavior, different from traditional ones. In particular, results show that those approximate fault tolerance techniques are less costly than traditional ones and can achieve almost the same reliability level.

The scientific works, experiments, and results presented in this book were developed over around five to six years of research. The main author completed his Ph.D. and acted as a researcher on a number of institutes and projects during that time.

Porto Alegre, Brazil — Gennaro S. Rodrigues
Porto Alegre, Brazil — Fernanda L. Kastensmidt
Lyon, France — Alberto Bosio

Contents

Part I Motivation and Introduction

1 Introduction 3
1.1 Embedded Systems and Approximation 3
1.2 Safety-critical Systems 4
1.3 Fault Tolerance 5
1.4 This Book 7
References 8

2 An Introduction to the Approximate Computing Paradigm 11
2.1 Summary 11
2.2 Quality Metrics 12
2.3 Approximation Methods 14
2.3.1 Types of Approximation 14
2.3.2 Technological Implementations 17
2.4 Conclusion 19
References 20

Part II Radiation Effects and Evaluation Methodologies

3 Radiation Effects on Digital Devices 25
3.1 Summary 25
3.2 Radiation Environment 25
3.3 Radiation Effects 27
3.3.1 Single Event Effects (SEE) 28
3.3.2 Total Ionizing Dose (TID) 32
3.3.3 Displacement Damage (DD) 33
3.4 Radiation-Induced Soft Errors in Zynq-7000 FPGA 33
3.5 Conclusion 34
References 35

4 Methodologies for Testing and Assessing Electronic and Computing Systems 37
4.1 Summary 37
4.2 Fault Tolerance Metrics 38
4.3 Onboard Fault Injection Emulation on FPGA 39
4.4 Onboard Fault Injection Emulation on Embedded Processor 40
4.5 Fault Injection Simulation 43
4.6 Laser Fault Injection 45
4.7 Conclusion 48
References 48

Part III Fault Tolerance and Approximation in Practice

5 Embedded Systems Fault Tolerance 53
5.1 Summary 53
5.2 Fault Tolerance 53
5.3 Reliability of Parallel Embedded Software on Multicore Processors 56
5.4 Parallel Fault Tolerance 64
5.5 Conclusion 67
References 68

6 Approximate Computing and Fault Tolerance 71
6.1 Summary 71
6.2 Approximation Methods 71
6.2.1 Data Precision Reduction 72
6.2.2 Successive Approximation 73
6.2.3 Taylor Series Approximation 75
6.3 Approximate Fault Tolerance: Discussion and Motivations 79
6.4 Approximate Triple Modular Redundancy (ATMR) 81
6.4.1 Hardware ATMR Based on Data Precision Approximation 81
6.4.2 Software ATMR Based on Successive Approximation and Loop-Perforation 85
6.4.3 Parallel Software ATMR Based on Function Skipping 87
6.5 Parallel Approximate Error Detection (PAED) 89
6.6 Conclusion 91
References 92

7 Experimental Analysis and Discussion 95
7.1 Summary 95
7.2 Approximation Methods 96
7.2.1 Taylor Series Approximation 96
7.2.2 Successive Approximation and Loop-Perforation 100
7.2.3 Behavior and Application Evaluation on Operating Systems 110

7.3 Hardware ATMR 114
7.3.1 Random Accumulated Fault Injection 115
7.3.2 Exhaustive Fault Injection 117
7.4 Software ATMR 118
7.5 PAED 123
References 127

8 Final Conclusions and Remarks 129
8.1 Summary 129
8.2 Approximate Computing 129
8.3 Safety-Critical Systems, Reliability, and Approximate Fault Tolerance 130
8.4 Experimental Results 130

Acronyms

ABFT	Application-Based Fault Tolerance
AMBA	ARM Advanced Microcontroller Bus Architecture
API	Application Programming Interface
ATMR	Approximate Triple Modular Redundancy
CDMR	Conditional Double Modular Redundancy
CMP	Chip Multiprocessor
COTS	Commercial Off-The-Shelf
CRC	Cyclic Redundancy Check
DD	Displacement Damage
DUT	Design Under Test
DVFS	Dynamic Voltage and Frequency Scaling
DWC	Duplication With Comparison
ECC	Error Correction Code
EDDI	Error Detection by Duplicated Instructions
FFT	Fast Fourier Transform
FI	Functional Interruption
FIM	Fault Injection Module
FIT	Failure in Time
FPGA	Field-Programmable Gate Array
HDL	Hardware Description Language
HLS	High-Level Synthesis
HPC	High-Performance Computing
IC	Integrated Circuit
ICAP	Internal Configuration Access Port
IP	Intellectual Property
LET	Linear Energy Transfer
MBU	Multibit Upset
MCU	Multicell Upset
MSE	Mean Square Error
MWTF	Mean Work To Fail

NIEL	Non-Ionizing Energy Loss
NVP	N-Version Programming
OCM	On-Chip Memory
OpenMP	Open Multi-Processing
OS	Operating System
OVPSim	OVP Simulator
PAED	Parallel Approximate Error Detection
PL	Programmable Logic
PS	Processing System
PSNR	Peak Signal-to-Noise Ratio
ROI	Region of Interest
RTCA	Radio Technical Commission for Aeronautics
SDC	Silent Data Corruption
SEB	Single Event Burnout
SEE	Single Event Effect
SEFI	Single Event Functional Interrupt
SEGR	Single Event Gate Rupture
SEL	Single Event Latch-up
SER	Soft Error Rate
SET	Single Event Transient
SEU	Single Event Upset
SHE	Single Hard Error
SIHFT	Software-Implemented Hardware Fault Tolerance
SoC	System-on-a-chip
TID	Total Ionizing Dose
TMR	Triple Modular Redundancy
TPA	Two-Photon Absorption
UT	Unexpected Termination
μSEL	Micro Single Event Latch-Up

Part I
Motivation and Introduction

Introduction 1

1.1 Embedded Systems and Approximation

Factors such as power efficiency and execution performance are of great importance for embedded systems and can be improved through approximate computing [1]. The approximate computing paradigm works with the idea that most applications are able to tolerate some flexibility (i.e., degradation) in the computed result based on a quality threshold specified by the application requirements. Indeed, several algorithms can present a *"good enough"* result even when they are executed on inexact computation units. An image processing algorithm, for example, might be able to tolerate some variations in the output quality, given the fact that the human eye is not able to perceive minor differences between images. Such an algorithm might therefore skip some computation in order to execute faster and have a lower memory footprint, causing an acceptable degradation of the output image. In that context, approximate computing has been proposed as a means to provide computational resources savings, alongside execution time and energy consumption reduction, with controlled quality degradation. Approximate computing techniques can be applied at every level of the computational stack, from circuits and hardware to software. Those techniques [2] have been used in many scenarios, from big data, scientific applications to embedded systems [3].

The literature presents a plethora of approximation strategies, both for software and hardware. The loop-perforation technique is an excellent software approximation example, being able to achieve useful outputs while not executing all the iterations of an iterative code [4]. Indeed, authors claim this approach typically delivers performance increases of over a factor of two while introducing an output deviation of less than 10%. Another approximation technique for software applications consists of reducing the bit-width used for data representation [5]. This technique also achieves a better execution speed than its non-approximate counterparts. Hardware-based approximation techniques usually make use of alternative speculative implementations of arithmetic operators. An example of this approach is the

G. S. Rodrigues et al., *Approximate Computing and its Impact on Accuracy, Reliability and Fault-Tolerance*, Synthesis Lectures on Engineering, Science, and Technology,
https://doi.org/10.1007/978-3-031-15717-2_1

implementation of variable approximation modes on operators [6]. Hardware approximation is also present in the image processing domain in the form of approximate compressors [7].

The quality degradation inherent to approximate computing is not to be forgotten. Although some quality degradation is acceptable for image processing algorithms (as exemplified before), it might not be acceptable for high dependability systems. A 10% quality degradation on an image might pass by unperceived, but an error of 10% in a banking system that computes the profits and taxes from a conglomerate will indeed be a severe problem. Even the perfect example of acceptable quality degradation (image processing) calls for an in-depth analysis of how acceptable that is: surely no graphics processing unit manufacturer wants to be known as the one which provides low-quality graphics at a very fast frame rate or vice versa.

1.2 Safety-critical Systems

Safety-critical applications are an excellent example of a category in which approximate computing can indeed bring good fruits, but it requires careful implementation. Applications defined as safety-critical deal with human lives and high-cost equipment and therefore require high dependability, i.e., low error rates. Safety-critical systems such as aerospace and avionics applications are often exposed to space radiation. Indeed, even systems that operate at ground level can be subject to space radiation [8], and some of those are also categorized as safety-critical systems (e.g., self-driven cars and their collision avoidance algorithms).

Radiation effects in semiconductor devices vary from transient data corruption to permanent damage. The state of a memory cell, register, latch, or any part of the circuit that holds data can be indeed changed by a radiation event. Single radiation events might cause *soft* or *hard* errors. Soft errors are the primary concern for commercial applications and can manifest themselves in many ways [9]. They occur when a radiation event is strong enough to change the data state without permanently damaging the system [10]. In software applications, those errors can be categorized into two major groups: silent data corruption (SDCs) and functional interruption (FIs) [11]. An SDC occurs when the application finishes properly, but its final memory state differs from the expected gold state. FIs are considered when the application hangs or terminates unexpectedly. From the user point of view, SDCs are the most critical cases since there is no information provided by the system to assess the output quality. On the other hand, FIs are immediately detected by the user that can thus take appropriate counter measures (e.g., reset the system). Please note that the "user" in this context can be a human being or another computing system. Hard errors are permanent damages to the system and are often related to dose-rate radiation effects (i.e., associated with the accumulation of radiation and its impacts on the behavior of the transistor).

The new transistor technologies have reduced their dimensions and operation thresholds and thus improved their energy consumption and performance. Their sensitivity to radiation, however, is often not a concern for the industry that focuses its efforts on higher transistor density and functionality at a low cost. Indeed, the reduction of the transistor size can now make them the target of radiation-induced faults that would otherwise occur in space environments at ground level [12]. Although those fault-induction-related issues are not a significant concern for the traditional consumer, which can accept sparse minor errors, they are indeed a severe concern for safety-critical systems.

1.3 Fault Tolerance

The traditional hardware manufacturers are not motivated to develop new radiation-hardened technologies because of their high development cost and, consequently, low-profit margin due to limited applications [8]. On the other side, the safety-critical industry is also often not interested in radiation-hardened hardware, which is expensive and does not provide the same performance as state-of-the-art hardware devices. The industry has thus turned to commercial off-the-shelf (COTS) embedded processors, and systems-on-a-chip (SoC) combined with fault tolerance techniques [13]. COTS are typically low-cost, very flexible, and consume little power. However, COTS do not provide inherent fault tolerance (apart from traditionally used methods such as memory error correction codes, which alone do not provide all the reliability required by safety-critical systems). They, therefore, call for hardware- and software-based fault tolerance techniques to assure reliability. The Zynq-7000 All Programmable SoC [14], for instance, is an example of a COTS system composed of two ARM processor cores and a field-programmable gate array (FPGA) that is capable of serving a wide range of safety-critical applications (that is, if fault tolerance methods are good enough to support the claim). COTS devices also provide a myriad of system configuration parameters, which may directly affect fault tolerance.

Fault tolerance can be applied at the hardware level by duplicating or triplicating a component and adding voters and checkers that verify the consistency of the processed data. Those techniques, however, introduce a prohibitive area and power overhead. Software-based fault tolerance techniques do not need extra hardware and are widely presented and discussed in the literature [15, 16]. In that case, redundancy is applied at the task level and executed in single or multiple processing cores. Although software-based techniques may present no hardware area overhead, they pay the cost of execution time and memory footprint, as well as energy consumption (that derive from those). One example of a fault tolerance mechanism that can be applied to both hardware and software is duplication with comparison (DWC), which duplicates the application and implements a checker to compare any discrepancy between the data generated by the two independent executions. DWC is capable of finding errors but not correcting (i.e., masking) them. A third execution would be needed to mask the error by voting for the correct data.

Concerning fault tolerance, approximate computing can mask a higher number of errors by relaxing data precision requirements. On systems that do not need high accuracy or quality, the approximation can be used because the small errors it introduces are not big enough to be considered a problem. Besides, the execution time reduction attained by approximate computing can improve application reliability by reducing its exposure time: an application that executes faster will be subject to less radiation and, therefore, fewer radiation-induced faults. SoCs arise as perfect implementation platforms for approximate computing. Industry-leading companies offer SoC presenting both an FPGA logic layer (PL) and an embedded processor as a processing system (PS), such as the aforementioned Zynq-7000 All Programmable SoC. Approximate computing projects can profit from the hardware-software co-design made available from COTS systems to implement any level of approximation or as a means of co-processing. Indeed, as this book will further detail, many approximation techniques consist of executing approximate versions of standard functions on programmable hardware instead of hardcore microprocessors.

In the literature, several works investigated the use of approximate computing in the context of fault tolerance. This research area is generally known as Approximate Triple Modular Redundancy (ATMR) [17] applied at the circuit level. The ATMR approach employs three Approximate Integrated Circuits (AxICs) instead of three fully precise replicas. For a given input, only one AxIC can give an incorrect answer. However, ATMR fault tolerance capability mainly depends on the voter. Let us resort to an example to illustrate this issue. Let X be an input vector for the ATMR replicas. Let one of the three AxICs produce a wrong response due to the approximation, while the other two produce a correct response. Let us imagine that a soft error occurs in one of the AxICs, thus modifying its output. If it occurs in the AxIC providing the approximate output, then the voter will still be able to produce the correct response, thanks to the two remaining AxICs. Conversely, if an AxIC providing the correct output to X experiences the error, there will be two incorrect responses, i.e., the approximate one and the faulty one. Thus, the voter may likely produce a wrong response. In summary, input vectors for which only two out of three AxICs compute correctly are not protected against faults. Recently, a novel approach referred to as Quadruple Approximate Modular Redundancy (QAMR) was presented [18, 19]. The QAMR is based on the idea of approximating a subset of the circuit selectively. The final goal is to have four approximate replicas in such a way as to guarantee that, among the four, at least three of them will provide precise results for a given output. The benefit is that the voter does not have to be modified. However, the cost of design is higher since the approximation has to be carefully identified. The above works targeted fault tolerance through masking (TMR). Concerning duplication with comparison, we must cite [20] where authors proposed using approximate computing for image processing applications DWC. The main concept is to divide the application into a set of "filters". Each filter applies a computation and the output of $filter_i$ is the input of $filer_{i+1}$. For each filter, an approximation replica is implemented based on the use of neural network approximation [21]. Then, for each filter, an approximate checker is designed to compare the approximate and the precise outputs.

1.4 This Book

This book presents the bases of the possible usage of approximate computing on safety-critical systems. The approximate computing paradigm can achieve several fundamental requirements of embedded safety-critical systems, such as low power consumption and high performance. Those, however, are achieved at the cost of precision and accuracy loss, which are severe concerns for critical applications. Another significant issue is reliability: approximate safety-critical systems shall be able to tolerate errors or at least support traditional fault tolerance techniques. Therefore, it is essential to know not only the improvements approximate computing can bring to a project but also its costs and how it affects its dependability.

We present techniques applied at the hardware and software level, as well as discussions on the implementation costs and precision impact of approximation and fault tolerance on the system. The approximation methods introduced in the book are also subject to fault tolerance analysis by fault injection experiments. Those experiments are intended to evaluate both the inherent fault tolerance of approximate computing and the efficacy of traditional fault tolerance mechanisms when applied to an approximate application. The book further introduces novel approximate fault tolerance techniques, based on the traditional ones presented by the literature, intended to provide fault coverage close to their non-approximate counterparts but at lower costs.

As discussed in the sections above, safety-critical systems often use fault tolerance techniques to assure dependability. Those fault tolerance methods, however, can themselves also be approximated. Approximating fault tolerance techniques might reduce their computational costs and, in some cases, even make them arguably more reliable. For example: by reducing the memory usage or the hardware used size of an FPGA-implemented system, the probability of a radioactive particle hitting the electronic system in a critical area is thus also reduced. This book approaches the approximate fault tolerance techniques question by evaluating, comparing, and discussing those techniques between their traditional and approximated counterparts.

The book thus presents and evaluates

- Approximate computing methods,
- Fault tolerance techniques, and
- Approximate fault tolerance techniques.

Also, four different fault injection and fault tolerance evaluation methodologies are presented:

- Fault injection emulation,
- Fault injection simulation,
- Laser radiation experiments, and
- Heavy-ion radiation experiments.

Each one of those methodologies serves better for a specific evaluation purpose. The fault injection emulation on programmable hardware, for example, might be used to evaluate the behavior of the design under a situation of cumulative faults in an effort to find out at which point (given the number of accumulated injected faults) the design begins to present errors. It can also be used to perform exhaustive studies on programmable hardware to determine which bits of the bitstream used to program the FPGA are critical (i.e., a bit-flip on this bit will provoke errors). On the other hand, fault injection simulation can inject faults on the register file of the processor to analyze which are the most critical registers and how faults affecting the register file propagate to become errors in a given context.

The question of whether approximate computing by itself is safe or not to be used as the primary development strategy for safety-critical systems is complex, and its answer most probably relies on the particularities of each project. The same is arguably true for the idea of approximating fault tolerance techniques. The works presented in this book intend to shine a light on that question and move it forward to a definitive answer.

References

1. J. Han, M. Orshansky, Approximate computing: an emerging paradigm for energy-efficient design, in *2013 18th IEEE European Test Symposium (ETS)* (2013), pp. 1–6
2. S. Venkataramani, S.T. Chakradhar, K. Roy, A. Raghunathan, Approximate computing and the quest for computing efficiency, in *Proceedings of the 52Nd Annual Design Automation Conference, DAC '15* (ACM, New York, NY, USA, 2015), pp. 120:1–120:6
3. R. Nair, Big data needs approximate computing: technical perspective. Commun. ACM **58**(1), 104–104 (2014). (December)
4. S. Sidiroglou-Douskos, S. Misailovic, H. Hoffmann, M. Rinard, Managing performance vs. accuracy trade-offs with loop perforation, in *Proceedings of the 19th ACM SIGSOFT Symposium and the 13th European Conference on Foundations of Software Engineering, ESEC/FSE '11* (ACM, New York, NY, USA, 2011), pp. 124–134
5. C. Rubio-Gonzalez, C. Nguyen, H.D. Nguyen, J. Demmel, W. Kahan, K. Sen, D.H. Bailey, C. Iancu, D. Hough, Precimonious: tuning assistant for floating-point precision, in *Proceedings of the International Conference on High Performance Computing, Networking, Storage and Analysis, SC '13* (ACM, New York, NY, USA, 2013), pp 27:1–27:12
6. M. Shafique, W. Ahmad, R. Hafiz, J. Henkel, A low latency generic accuracy configurable adder, in *Proceedings of the 52Nd Annual Design Automation Conference, DAC '15* (ACM, New York, NY, USA, 2015), pp. 86:1–86:6

7. A. Momeni, J. Han, P. Montuschi, F. Lombardi, Design and analysis of approximate compressors for multiplication. IEEE Trans. Comput. **64**(4), 984–994 (2015). (April)
8. R.C. Baumann, Radiation-induced soft errors in advanced semiconductor technologies. IEEE Trans. Device Mater. Reliab. **5**(3), 305–316 (2005). (Sept)
9. C. Poivey, J.L. Barth, K.A. LaBel, G. Gee, H. Safren, In-flight observations of long-term single-event effect (SEE) performance on orbview-2 solid state recorders (SSR), in *2003 IEEE Radiation Effects Data Workshop* (2003), pp. 102–107
10. A.J. Tylka, W.F. Dietrich, P.R. Boberg, E.C. Smith, J.H. Adams, Single event upsets caused by solar energetic heavy ions. IEEE Trans. Nucl. Sci. **43**(6), 2758–2766 (1996)
11. M.-C. Hsueh, T. K. Tsai, R. K. Iyer, Fault injection techniques and tools. Computer **30**(4), 75–82 (1997)
12. J. Tausch, D. Sleeter, D. Radaelli, H. Puchner, Neutron induced micro SEL events in cots SRAM devices, in *2007 IEEE Radiation Effects Data Workshop* (2007), pp. 185–188
13. M. Pignol, Cots-based applications in space avionics, in *2010 Design, Automation Test in Europe Conference Exhibition (DATE 2010)* (2010), pp. 1213–1219
14. M. Al Kadi, P. Rudolph, D. Gohringer, M. Hubner, Dynamic and partial reconfiguration of Zynq 7000 under Linux, in *2013 International Conference on Reconfigurable Computing and FPGAs (ReConFig)* (IEEE, 2013), pp. 1–5
15. G.K. Saha, Software based fault tolerance: a survey. Ubiquity 1:1–1:1 (2006)
16. L. Osinski, T. Langer, J. Mottok, A survey of fault tolerance approaches on different architecture levels, in *ARCS 2017; 30th International Conference on Architecture of Computing Systems* (2017), pp. 1–9
17. B.D. Sierawski, B.L. Bhuva, L.W. Massengill, Reducing soft error rate in logic circuits through approximate logic functions. IEEE Trans. Nucl. Sci. **53**(6), 3417–3421 (2006). (December)
18. B. Deveautour, M. Traiola, A. Virazel, P. Girard. QAMR: an approximation-based fully reliable TMR alternative for area overhead reduction, in *2020 IEEE European Test Symposium (ETS)*. (IEEE, 2020)
19. M. Traiola, J. Echavarria, A. Bosio, J. Teich, I. O'Connor, Design space exploration of approximation-based quadruple modular redundancy circuits, in *2021 IEEE/ACM International Conference On Computer Aided Design (ICCAD)* (IEEE, 2021)
20. M. Biasielli, C. Bolchini, L. Cassano, A. Mazzeo, A. Miele, Approximation-based fault tolerance in image processing applications. IEEE Trans. Emerg. Top. Comput. 1–1 (2021)
21. S. Barone, M. Traiola, M. Barbareschi, A. Bosio, Multi-objective application-driven approximate design method. IEEE Access **9**, 86975–86993 (2021)

An Introduction to the Approximate Computing Paradigm

2

2.1 Summary

Approximate computing has been proposed as an approach for developing energy-efficient systems [2], lowering hardware resource usage and presenting better execution times, and has been used in many scenarios, from big data to scientific applications [15]. It can be achieved in a multitude of ways, ranging from transistor-level design to software implementations. Many systems do not take precision and accuracy as essential assets. Those are the ones that can profit from this computational paradigm [29]. An example of this type of application is image processing, where the degradation of some pixels will not impact the overall image quality. Moreover, the approximation can also be unavoidable. Floating-point operations, for instance, have frequent rounding of values, making them inherently approximate. Numerical algorithms are also frequently of approximate nature: the calculation of an integral, for example, is not natural for a computer and consists of an iterative calculation of a finite sum of terms (and not an infinite one, as the mathematical theory defines it).

Even in systems where quality and accuracy are essential, the mere definition of a good quality result can be malleable. In image processing, for example, the final output is evaluated from a human perspective (the quality of the image). This perspective is subjective: some people are more capable of analyzing the quality of an image than others, and the definition of a "good enough" quality is even more debatable. Typical error-resilient image processing algorithms can indeed accept errors of up to 10% [17], which would be unacceptable for a military system calculating the trajectory of a ballistic projectile, for example. This margin of error acceptance can be exploited to improve energy consumption and execution performance.

The weak definition of "error acceptance" can also be used by approximate computing for quality configuration. Given that different systems have different quality requirements, a designer might use just the necessary energy, hardware area, or execution time to meet the

G. S. Rodrigues et al., *Approximate Computing and its Impact on Accuracy, Reliability and Fault-Tolerance*, Synthesis Lectures on Engineering, Science, and Technology,
https://doi.org/10.1007/978-3-031-15717-2_2

needs of his project. An excellent example of how a circuit can be configured in that manner is by using different refresh rates for memories [4], or different precision for data storage and representations [25]. The image processing domain is particularly interesting because it exemplifies how approximation can be implemented on different levels. For example, a minor loss in precision can be accepted by applying approximation via hardware, by reducing the refresh rates of eDRAM/DRAM and the SRAM supply voltage, reducing energy consumption [13]. On a higher level of implementation, the approximation can be used by loop-perforation (finishing a loop execution earlier than expected) [23] or by executing specific functions on neural accelerators [14].

This chapter will overview the most used approximation methods applied at both software and hardware levels. As will be further described, some approximation ideas can actually be applied in more than one computing stack. The chapter is organized as follows: Sect. 2.2 presents the quality metrics (i.e., precision and accuracy evaluation) used for different applications when dealing with approximation. Section 2.3 presents approximation methods applied to different techniques at software, architectural, and hardware levels. Finally, we review some of the state-of-the-art implementations of the previously discussed techniques in Sect. 2.3.2.

2.2 Quality Metrics

Given the plethora of approximation methods and systems that make use of them, the literature also presents an extensive list of error metric definitions. Some examples of how precision loss is measured on approximate systems are

- **Pixel Difference:** It consists of a complete comparison of two images pixel-by-pixel, where every pixel is represented by a value. It is customarily used to compare grayscale images, where the pixel value defines the gray intensity (the higher value being complete darkness).
- **Peak signal-to-noise ratio (PSNR):** It is calculated using the mean square error (MSE) between the two images (the original and the approximate one). It indicates the ratio of the maximum pixel intensity to the distortion. It is calculated by the formula $\text{PSNR} = 10log_{10}(\text{MAX}^2/\text{MSE})$, where MAX stands for the maximum value of a pixel in the images.
- **Hamming Distance:** By comparing data in a bitwise manner, the hamming distance consists of the number of positions where the bit values are different between two binary strings.
- **Ranking Accuracy:** When approximate computing is applied to ranking algorithms, such as the ones used by search engines, it can generate different results depending on the ranking definitions and the algorithms used. A research result from Bing and Google

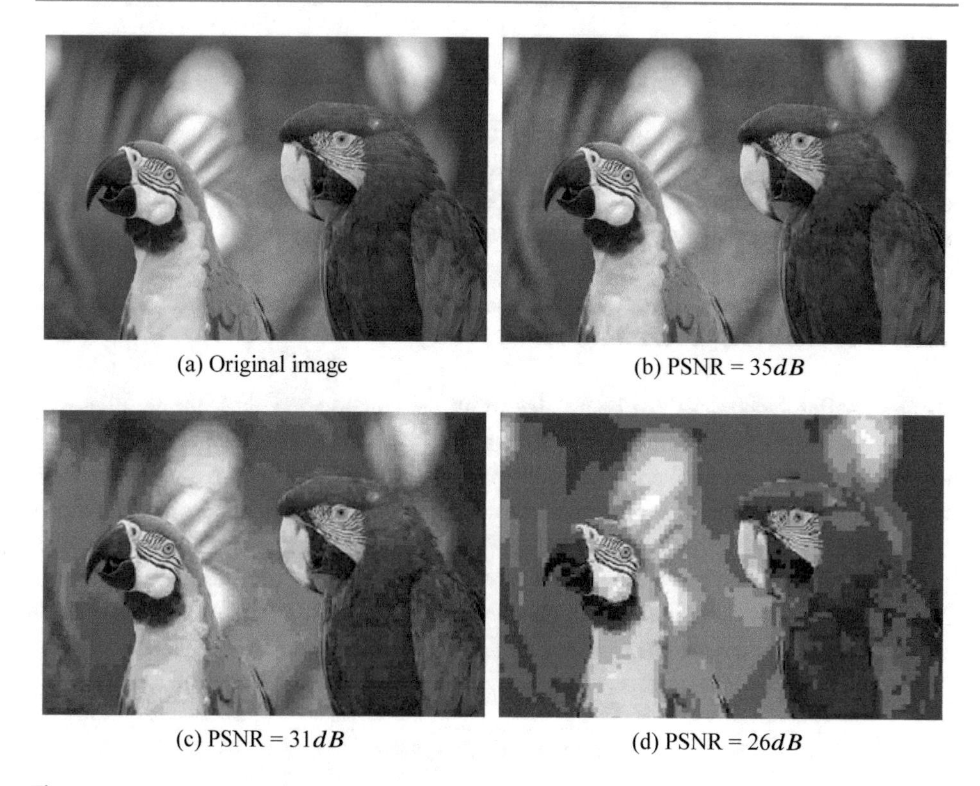

(a) Original image (b) PSNR = $35dB$

(c) PSNR = $31dB$ (d) PSNR = $26dB$

Fig. 2.1 Examples of the effect of different values of PSNR on image quality

search engines, for example, will likely be different. The accuracy is defined based on pre-established parameters.

- **Error Probability:** It consists of comparing all possible outputs of an approximate function and its non-approximate counterpart. This metric provides a probability value for an approximation to present an error. However, it does not evaluate the criticality or impact of that error.
- **Relative Difference:** presents the error in relation to the standard output. This metric is capable of evaluating the criticality of an error (Fig. 2.1).

The presented quality evaluation metrics are not mutually exclusive. One application might use several different parameters to evaluate its quality loss. Both PSNR and pixel difference are used as image quality metrics, for example.

2.3 Approximation Methods

It is not easy to define approximation by itself, especially given the multiple ways it can be implemented for many applications. In the remainder of this book, several approximation techniques will be presented and evaluated. For now, this chapter will focus on detailing all the approximation capabilities of the multiple computing stacks.

2.3.1 Types of Approximation

Approximation techniques can be applied to all the computation stack levels. Figure 2.2 divides approximation techniques into three groups that define their implementation level: software, architectural, and hardware levels. As Fig. 2.2 shows, some approximation methods can even be implemented on more than one level. Load value approximation, for example, can be implemented by both purely software approaches or in memory control units. Figure 2.2 presents only some of the most used approximation methods and the most discussed in the literature. However, there are uncountable ways of approximating an application, and the very definition of what is to be considered an approximation or not is debatable.

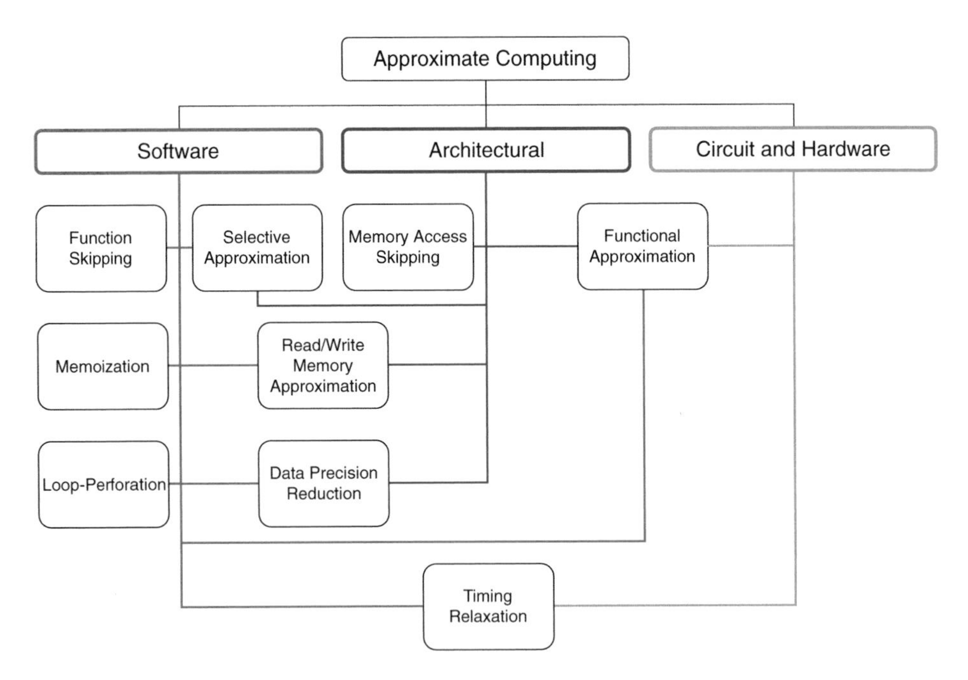

Fig. 2.2 Approximate computing classification. *Source* Author

The techniques presented in Fig. 2.2 are defined below.

- **Functional Approximation:** It consists of implementing approximate versions of an algorithm. The literature presents many functional approximation techniques for circuits and architecture levels that implement approximate arithmetic operators (e.g., adders and multipliers). One approach is to remove the carry chain from the circuit to reduce delay and energy consumption, which can be done by altering the subedars of a standard adder cell of n bits [8]. In [10], the authors presented an approximate 2×2 multiplier design that gives correct outputs for 15 of the 16 possible input combinations and uses half of the area of a standard non-approximate multiplier. Functional approximation can be implemented using neural networks at the software and architectural levels. A neural network can learn how a standard function implementation behaves concerning different inputs via machine learning. In a complex system, the neural network can be used to implement approximate functions via software-hardware co-design. Traditional approximable codes can be transformed into equivalent neural networks with lower output accuracy but better execution time performance [1].
- **Selective Approximation:** Some systems are made of parts that don't need to provide accuracy as much as others. The idea is to take advantage of the fact that even inside an algorithm, some parts affect more the final accuracy than others. Those parts can be approximated to provide energy consumption reduction and improve execution time performance [26]. On the architectural level, selective approximation is implemented as alternative speculative arithmetic operators. An example of this approach is the implementation of variable approximation modes on operators [22]. When applied to software applications, approximate computing usually consists of inexact computations, which provide results with lower accuracy than usual [27]. Most approximate computing techniques for software consist of modifying the algorithm so that it executes approximately, providing a final result more rapidly.
- **Function Skipping:** In a system composed of tasks that complement each other in the sense of providing a final result, some of the tasks can be skipped while maintaining a level of accuracy and error resiliency defined by the user [31].
- **Memoization:** Traditionally, memoization consists of saving outputs of functions for given inputs to be reused later. Given that some input data are frequently reused, their calculated outputs can be stored and used without the need to re-execute the function. If *similar* inputs provide *similar* results for a given function, it means that an already-calculated function output can approximately cover a range of inputs. In [9], the authors propose approximate value reuse for memoization, providing very low accuracy loss.
- **Loop-Perforation:** In loop-based algorithms, loop-perforation is used to reduce execution time. An excellent example is numerical algorithms. The calculation of an integral using the trapezoidal method, for example, consists of calculating the area of a high number of trapezoids under the curve of a function, providing an approximation of the area beneath it. By reducing the number of calculated trapezoids, the final value will

be less accurate, but the program will finish earlier. The literature also presents techniques to apply this approximation method to general-use algorithms, filtering out the loops that cannot be approximated and using loop-perforation on those that can [23]. The authors claim this approach typically delivers performance increases of over a factor of two while introducing an output deviation of less than 10%. However, loop-perforation is an algorithm-based approximate technique, as it is only applicable to loop-based code, limiting its applicability.

- **Data Precision Reduction:** Data precision reduction is one of the techniques that can be implemented both at the software and architectural levels. Reducing the data precision of an application (i.e., the number of bits used to represent the data) is a straightforward technique to reduce memory footprint. Reducing memory usage also reduces energy consumption at the cost of accuracy loss. In [7], the authors show that reducing floating-point precision on mobile GPUs can reduce energy consumption with image quality degradation. This degradation, however, can be acceptable and even unperceivable for the human eye. Lower memory utilization is suitable for safety-critical systems because it reduces the essential and critical bit count, making them less susceptible to faults. Reducing the bit-width used for data representation is also a popular approximation method [20].
- **Timing Relaxation:** Operating voltage can be scaled at the circuit level, impacting the effort expended on the computation of processing blocks inside the clock period. It affects the accuracy of the final result and also the energy consumption [3]. In [3], the authors propose the voltage over scaling of individual computation blocks, assuring that the accuracy of the results will "gracefully" scale with it. Voltage scaling can be implemented in hardware dynamically. Dynamic voltage and frequency scaling (DVFS), for example, is a power management technique used to improve power efficiency, reducing the clock frequency and the supply voltage of the processor [11]. However, DVFS can cause data cells to be stuck with a specific value because it reduces the threshold between a logical one and zero. This type of approximation impacts the integrity of the hardware and the precision of the data. Timing relaxation can also be implemented in software. On parallel programs, it is achieved by relaxing the synchronization between execution tasks [12].
- **Read/Write Memory Approximation:** It consists of approximating data that is loaded from or written in the memory or the read/write operations themselves. This is primarily used on video and image applications, for example, where accuracy and quality can often be relaxed, to reduce memory operations [6, 18]. In [25], the authors propose a dynamic bit-width technique based on the application accuracy requirements, where a control system determines the precision of data accesses and loads. The authors claim that it can be implemented in a general-processor architecture without the need for hardware modifications by communicating with off-chip memory via a software-based memory management unit. The approximation can also be applied to the cache memory. In the event of a load data cache miss, the processor must fetch the data from the following cache level or in the main memory. This can be a very time-consuming task. Load value

approximation can be used to estimate an approximate value instead of fetching the real one from memory. In [24], the authors present a technique that uses the GPU texture fetch units to generate approximate values. This approximation causes an error of less than 0.5% in the final image output while reducing the kernel execution time by 12%. In [21], the authors propose an approximation technique for multi-level cell STT-RAM memory technologies by lowering its reliability up to a user-defined accuracy loss acceptance. This memory technology has a considerable reliability overhead, which can be reduced. They selectively approximate the storage data of the application and reduce the error-protection hardware minimizing error consumption.

- **Memory Access Skipping:** Using a combination of the memoization and function skipping techniques, it is likewise possible to skip memory accesses. Non-critical data can be omitted as long as it will not heavily damage the output accuracy. Approximate neural networks can skip reading entire rows of their weight matrices as long as those neurons are not critical, reducing energy consumption and memory access and improving performance [30].

The presented approximation techniques can be implemented in a multitude of ways on various device levels with different impacts on the system behavior. Unfortunately, approximate computing at the software level is less presented in the literature than at the hardware level. This is probably due to the origins of approximation being on energy consumption reduction and neural network applications.

2.3.2 Technological Implementations

Approximation techniques are applied to all the computational levels, as showed Fig. 2.2. Likewise, they can be implemented in many abstraction levels. Selective approximation is an example of a technique that can be applied in the software and architectural computation stacks and implemented via software code modification, programmable hardware, and even circuit level. Loop-perforation can also be achieved via code modification for embedded software and programmable hardware (using HLS) or with HDL project modifications. The way the approximation techniques are technologically implemented also has an important impact on their performance.

One of the problems with selective and functional approximation is the induced amount of errors in the system output that is sometimes too high to be acceptable. Also, not all hardware approximate computing implementations incur hardware size reduction or energy gains. The works at the architectural level of approximate operators [22], for example, do not present a significant hardware implementation area reduction when compared to a traditional operator. Indeed, some of the approximate operators presented by [22] not only have no hardware area gains but take more area than conventional operators. Furthermore, the size of the used area on programmable hardware devices has a direct impact on system reliability [28]. Therefore, the quality loss (in this case manifested as errors in some operation results) introduced by the approximation would only be acceptable by safety-critical systems if its area is sharply reduced.

The timing relaxation through the voltage scaling technique can be applied at the processor architecture level and programmable hardware. At the architecture level, voltage scaling is implemented during the design of the circuit. Most FPGA manufacturers make the voltage scaling of the device possible through easy-to-use design tools. However, even though it will impact the performance of a software application, it is not part of the software approximation group because its implementation has no direct connection with software development.

Developing alternate approximate versions of an algorithm is very time-consuming and intellectually demanding work. Some works propose frameworks that identify approximable portions of code to deal with this issue. In [19], the authors present a framework to discover what the data are that can be approximated without significantly interfering with the output of the system. They do so by injecting faults in the variables and analyzing how they affect the quality of the output. Another method is to identify parts of an application code that can be executed on approximated hardware, saving resources and energy [5]. The approximation technique to be applied to the approximable parts of the application would depend on the application requirements. Although those frameworks are presented as general-use, the question remains if they really can be applied to every type of algorithm. Because they base their methodology on simulation, it is hard to believe that they are able to cover every possible kind of faults that can affect every system.

One of the techniques with the most straightforward implementation is data precision reduction. The way it can be used to approximate software and FPGA applications is obvious: it is a matter of code modification. In software, the precision of floating-point units can be easily modified with the use of dedicated libraries or even by merely changing the type of the variable. The same can be done at VHDL/Verilog projects: a design can be adapted to process smaller vectors of data. Data precision reduction can bring good improvements in area and energy costs for hardware projects but frequently does not present high cost reduction on software. Fixed-point arithmetic, for example, can be used to approximate mathematical functions, such as logarithm, on FPGA implementations providing low area usage [16]. On software, however, it can increase the execution time of the application because all the operations and data handling routines are implemented at the software level.

Similarly, the loop-perforation technique can be implemented both in software and programmable hardware code. The difference is that, on programmable hardware, a loop might be implemented either as many circuits executing in parallel (one being each iteration of the loop) or one circuit that is re-executed in a timeline. Therefore, the impact of loop-perforation on software and hardware implementations can be very different. On software, it will mainly impact the execution time of the application, while in an FPGA implementation, it could also affect the energy and area consumption.

One example of loop-perforation applied to a real case scenario can be seen in Fig. 2.4. This figure shows the output of a Sobel filter execution. The Sobel filter (sometimes referred

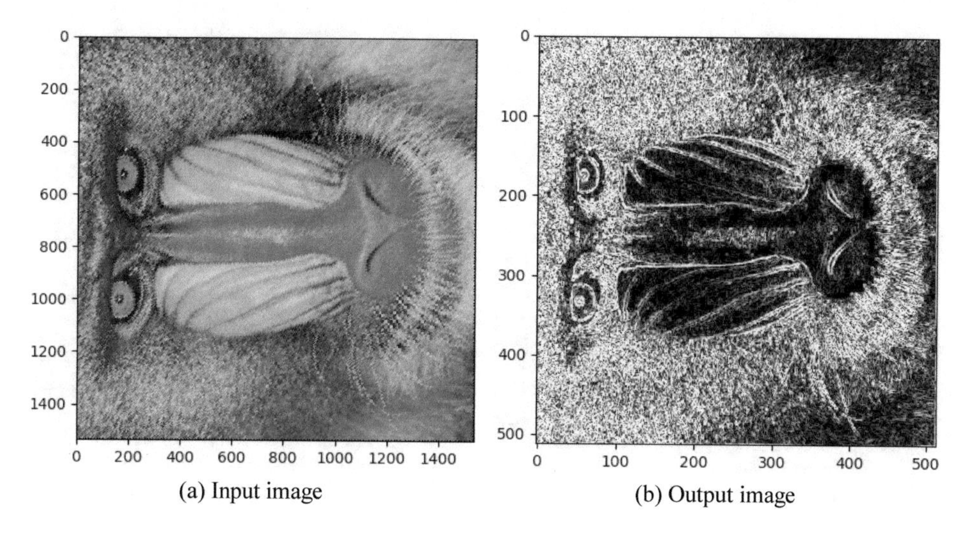

(a) Input image (b) Output image

Fig. 2.3 Sobel filter applied to example image

to as the Sobel operator) is an algorithm that, given an input image, generates an output emphasizing the original image's edges. Figure 2.3 shows an example of a Sobel filter input and output.

In Fig. 2.4, loop-perforation was applied to reduce the amount of processed pixels, thus reducing computation costs. The unprocessed pixels were filled by black (Fig. 2.4b), gray (Fig. 2.4c), and white (Fig. 2.4d). Notice that for each one of those cases, the MSE and SSIM values vary.

2.4 Conclusion

Approximation might significantly impact the overall output quality of a system. As we can see in Fig. 2.4, loop-perforation greatly impacts image processing if overused. This is especially true if, instead of only evaluating the metrics, we consider the perception of a human being (which is natural for systems in fields such as image and sound processing). However, as exemplified in Fig. 2.1, it is all about evaluating and tuning an acceptable threshold of quality degradation: some approximations might not even be noticeable by a human observer.

In the remainder of this book, we will apply the ideas developed in this chapter to a plethora of applications. Approximation will be applied to numerical and general-purpose algorithms as a primary means of development. It will also be studied as a tool for reducing computational costs and even hardware design costs. Finally, we will also explore its intrinsic fault tolerance and how it can be used for safety-critical systems to improve their performances not only on execution time and memory footprint but also on reliability.

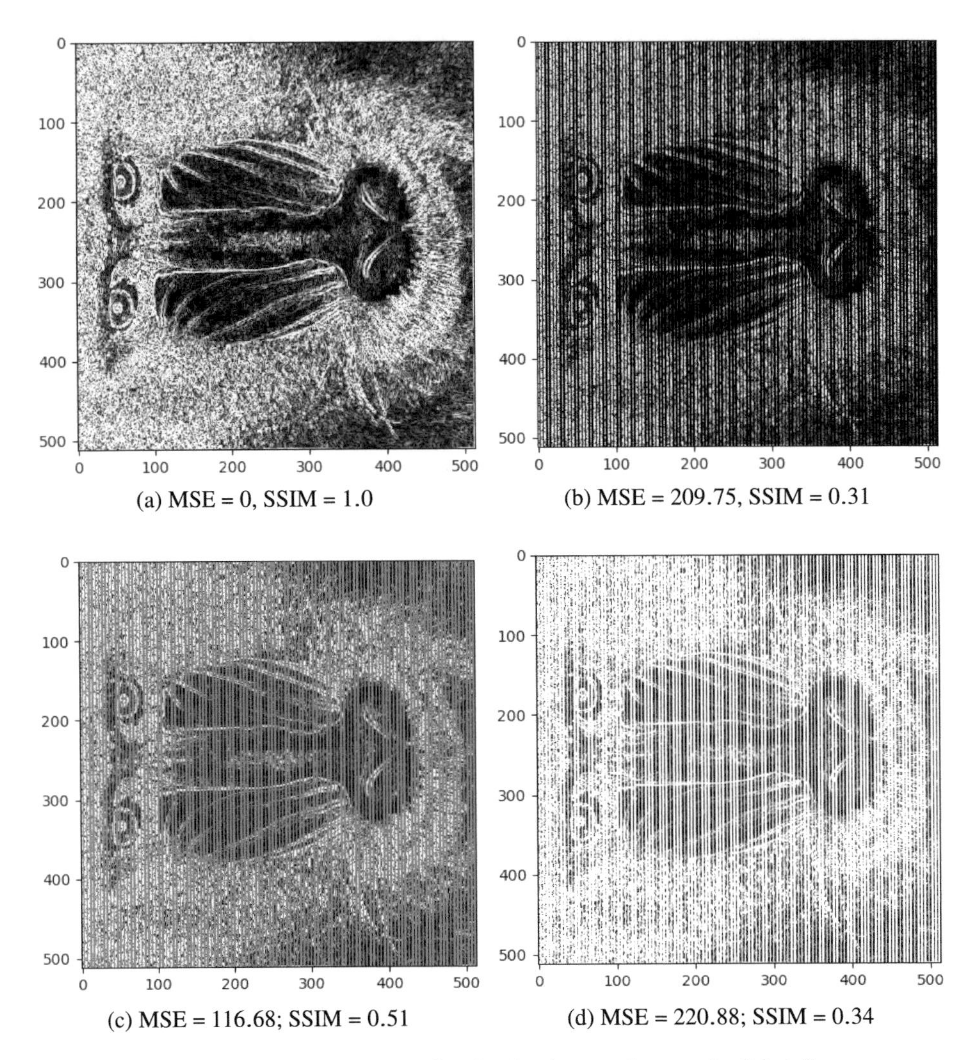

(a) MSE = 0, SSIM = 1.0

(b) MSE = 209.75, SSIM = 0.31

(c) MSE = 116.68; SSIM = 0.51

(d) MSE = 220.88; SSIM = 0.34

Fig. 2.4 Examples of the effect of loop-perforation implementations on the Sobel filter output

References

1. R. St. Amant, A. Yazdanbakhsh, J. Park, B. Thwaites, H. Esmaeilzadeh, A. Hassibi, L. Ceze, and D. Burger, General-purpose code acceleration with limited-precision analog computation, in *Proceedings - International Symposium on Computer Architecture*, pp. 505–516, (2014)
2. A. Bosio, D. Menard, O. Sentieys, *in Approximate Computing Techniques* (Springer, Berlin, 2022)

3. V.K. Chippa, D. Mohapatra, K. Roy, S.T. Chakradhar, A. Raghunathan, Scalable effort hardware design. IEEE Trans. Very Large Scale Integr. (VLSI) Syst. **22**(9), 2004–2016 (2014)
4. K. Cho, Y. Lee, Y.H. Oh, G. Hwang, J.W. Lee, eDRAM-based tiered-reliability memory with applications to low-power frame buffers, in *2014 IEEE/ACM International Symposium on Low Power Electronics and Design (ISLPED)* (2014), pp. 333–338
5. H. Esmaeilzadeh, A. Sampson, L. Ceze, D. Burger, Architecture support for disciplined approximate programming. SIGPLAN Not. **47**(4), 301–312 (2012)
6. Y. Fang, H. Li, X. Li, Softpcm: Enhancing energy efficiency and lifetime of phase change memory in video applications via approximate write, in *2012 IEEE 21st Asian Test Symposium* (2012), pp. 131–136
7. C.C. Hsiao, S.L. Chu, C.Y. Chen, Energy-aware hybrid precision selection framework for mobile GPUs. Comput. Graph. (Pergamon) **37**(5), 431–444 (2013)
8. A.B. Kahng, S. Kang, Accuracy-configurable adder for approximate arithmetic designs, in *DAC Design Automation Conference* (2012), pp. 820–825
9. G. Keramidas, C. Kokkala, I. Stamoulis, Clumsy value cache: an approximate memoization technique for mobile GPU fragment shaders, in *1st Workshop On Approximate Computing (WAPCO 2015)* (2015), p. 6
10. P. Kulkarni, P. Gupta, M. Ercegovac, Trading accuracy for power with an underdesigned multiplier architecture, in *2011 24th Internatioal Conference on VLSI Design* (2011), pp. 346–351
11. E. Le Sueur, G. Heiser, Dynamic voltage and frequency scaling: the laws of diminishing returns, in *Proceedings of the 2010 International Conference on Power Aware Computing and Systems, HotPower'10* (USENIX Association, Berkeley, CA, USA, 2010), pp. 1–8
12. S. Misailovic, D. Kim, M. Rinard, Parallelizing sequential programs with statistical accuracy tests. ACM Trans. Embed. Comput. Syst. **12**(2s) (2013)
13. S. Mittal, A survey of architectural techniques for improving cache power efficiency. Sustainable Computing: Informatics and Systems **4**(1), 33–43 (2014)
14. T. Moreau, M. Wyse, J. Nelson, A. Sampson, H. Esmaeilzadeh, L. Ceze, M. Oskin. Snnap: Approximate computing on programmable socs via neural acceleration, in *2015 IEEE 21st International Symposium on High Performance Computer Architecture (HPCA)* (2015), pp. 603–614
15. R. Nair, Big data needs approximate computing: technical perspective. Commun. ACM **58**(1), 104 (2014)
16. J.G. Pandey, A. Karmakar, C. Shekhar, and S. Gurunarayanan, An FPGA-based fixed-point architecture for binary logarithmic computation, in *2013 IEEE Second International Conference on Image Information Processing (ICIIP-2013)* (2013), pp. 383–388
17. A. Rahimi, A. Ghofrani, K. Cheng, L. Benini, R.K. Gupta, Approximate associative memristive memory for energy-efficient GPUs, in *2015 Design, Automation Test in Europe Conference Exhibition (DATE)* (2015), pp. 1497–1502
18. A. Ranjan, S. Venkataramani, X. Fong, K. Roy, A. Raghunathan, Approximate storage for energy efficient spintronic memories, in *2015 52nd ACM/EDAC/IEEE Design Automation Conference (DAC)*, pp. 1–6, (2015)
19. P. Roy, R. Ray, C. Wang, and Weng Fai Wong, Asac: automatic sensitivity analysis for approximate computing, in *Proceedings of the 2014 SIGPLAN/SIGBED Conference on Languages, Compilers and Tools for Embedded Systems, LCTES '14* (ACM, New York, NY, USA, 2014), pp. 95–104
20. C. Rubio-Gonzalez, C. Nguyen, H.D. Nguyen, J. Demmel, W. Kahan, K. Sen, D.H. Bailey, C. Iancu, D. Hough, Precimonious: tuning assistant for floating-point precision, in *Proceedings of the International Conference on High Performance Computing, Networking, Storage and Analysis, SC '13* (ACM, New York, NY, USA, 2013), pp. 27:1–27:12

21. F. Sampaio, M. Shafique, B. Zatt, S. Bampi, and J. Henkel, Approximation-aware multi-level cells stt-ram cache architecture, in *2015 International Conference on Compilers, Architecture and Synthesis for Embedded Systems (CASES)* (2015), pp. 79–88
22. M. Shafique, W. Ahmad, R. Hafiz, J. Henkel, A low latency generic accuracy configurable adder, in *Proceedings of the 52Nd Annual Design Automation Conference, DAC '15* (ACM, New York, NY, USA, 2015), pp. 86:1–86:6
23. S. Sidiroglou-Douskos, S. Misailovic, H. Hoffmann, M. Rinard, Managing performance vs. accuracy trade-offs with loop perforation, in *Proceedings of the 19th ACM SIGSOFT Symposium and the 13th European Conference on Foundations of Software Engineering, ESEC/FSE '11* (ACM, New York, NY, USA, 2011), pp. 124–134
24. M. Sutherland, J. San Miguel, and N.E. Jerger, Texture cache approximation on GPUs, in *Workshop on Approximate Computing* (2015), pp. 1–3
25. Y. Tian, Q. Zhang, T. Wang, F. Yuan, Q. Xu, Approxma: approximate memory access for dynamic precision scaling, in *Proceedings of the 25th Edition on Great Lakes Symposium on VLSI, GLSVLSI '15* (ACM, New York, NY, USA, 2015), pp. 337–342
26. V. Vassiliadis, K. Parasyris, C. Chalios, C.D. Antonopoulos, S. Lalis, N. Bellas, H. Vandierendonck, D.S. Nikolopoulos, A programming model and runtime system for significance-aware energy-efficient computing, in *Proceedings of the ACM SIGPLAN Symposium on Principles and Practice of Parallel Programming, PPOPP* (2015), pp. 275–276
27. S. Venkataramani, S.T. Chakradhar, K. Roy, A. Raghunathan, Approximate computing and the quest for computing efficiency, in *Proceedings of the 52Nd Annual Design Automation Conference, DAC '15* (ACM, New York, NY, USA, 2015), pp. 120:1–120:6
28. M. Wirthlin, High-reliability FPGA-based systems: Space, high-energy physics, and beyond. Proc. IEEE **103**(3), 379–389 (2015)
29. Q. Xu, T. Mytkowicz, N.S. Kim, Approximate computing: a survey. IEEE Des. Test **33**(1), 8–22 (2016)
30. Q. Zhang, T. Wang, Y. Tian, F. Yuan, Q. Xu, Approxann: an approximate computing framework for artificial neural network, in *2015 Design, Automation Test in Europe Conference Exhibition (DATE)* (2015), pp. 701–706
31. Í. Goiri, R. Bianchini, S. Nagarakatte, T.D. Nguyen, Approxhadoop: bringing approximations to mapreduce frameworks, in *Proceedings of the ACM International Conference on Architectural Support for Programming Languages and Operating Systems (ASPLOS)* (2015)

Part II
Radiation Effects and Evaluation Methodologies

Radiation Effects on Digital Devices

3

3.1 Summary

Safety-critical systems need to, by definition, be dependable. They might often put human lives at risk when they fail. Further, due to their practical application, they are subject to some forces that normal electronic systems are not (or are, but in a lower intensity). The biggest concern is regarding radiation. Avionics and military systems, for example, are subject to much higher levels of radiation than a personal computer. The same is obviously true for a satellite.

This chapter introduces and discusses those radiation effects and how they impact the functioning of digital devices. The chapter is structured as follows: Sect. 3.2 introduces the particularity of the space environment and its radiation; Sect. 3.3 presents the errors that are caused by this radiation and their differences. Then Sect. 3.4 proceeds to provide a practical example of how those errors might affect an FPGA. Finally, conclusions are drawn in the final Sect. 3.5.

3.2 Radiation Environment

Circuits operating in space and at higher altitudes are subject to a large spectrum of particles of different mass, each with a different energy range. The radiation experienced in space near Earth comes from several different sources, including protons and heavy ions from solar flares and cosmic rays. The space environment also includes protons and electrons trapped in the Van Allen Belts and heavy ions trapped in the Earth's magnetosphere, as shown in Fig. 3.1. Interactions of high-energy protons with a higher atmosphere can generate neutrons.

The interaction of those neutrons inside electronics can generate secondary ionizing particles, such as alpha particles, becoming an issue for electronics at avionics altitude.

G. S. Rodrigues et al., *Approximate Computing and its Impact on Accuracy, Reliability and Fault-Tolerance*, Synthesis Lectures on Engineering, Science, and Technology,
https://doi.org/10.1007/978-3-031-15717-2_3

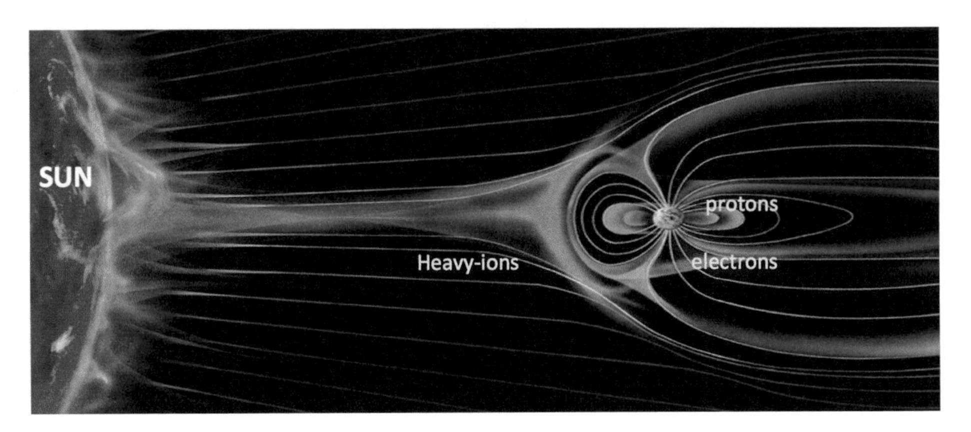

Fig. 3.1 Radiation effects in electronic devices. *Source* Author

Fig. 3.2 Radiation effects in electronic devices. *Source* Author

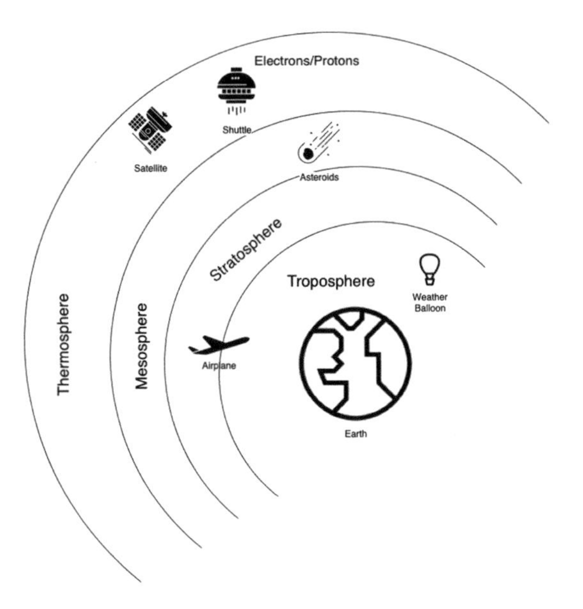

In a packaged semiconductor product, the primary source of alpha particles is the package materials, not the semiconductor device materials [3]. Most of these neutrons will decay in the atmosphere and reach ground level as thermal neutrons, but some high-energy neutrons can still contact electronics at ground level. Both thermal neutrons and high-energy neutrons can interact with electronics generating secondary ionizing particles, meaning that even circuits operating at sea level can experience transient faults caused by the interaction of low- and high-energy neutrons with the materials on the integrated circuit (Fig. 3.2).

Alpha particles are an important source of ionizing radiation and derive from the natural impurities in device materials. It is one of the radiations that can be emitted when the nucleus of unstable isotopes decays to a state of lower energy. Uranium and thorium are the most active radioactive isotopes and, therefore, the dominant source of alpha particles in materials alongside their daughter products. Their decay can produce alpha and beta particles, but the latter is not critical because they do not emit enough energy to cause ionization and provoke a soft error. Alpha particles induce electron-hole pairs in their awake and traveling in the silicon. In a packaged semiconductor product, the main source of alpha particles is the package materials, not the semiconductor device materials [3].

Neutrons from cosmic radiation with high energy can react with the silicon nuclei and produce secondary ions that induce soft errors. Indeed, cosmic radiation is one of the leading sources of errors in DRAM [14], and neutrons are the most likely cosmic radiation to cause soft errors in devices at ground level. Cosmic rays interact with the earth's atmosphere and produce cascades of secondary particles, which continue deeper, creating tertiary particles and so on. Less than 1% of the original flux reaches sea-level altitudes, and most of the particles that arrive consist of muons, pions, protons, and neutrons [19]. Muons and pions have a short life, and protons and electrons are attenuated by Coulombic interactions with the planet's atmosphere. Neutrons, however, have a higher flux and stability. Neutrons do not generate ionization in silicon alone. Instead, they interact with the chip materials breaking excited nuclei into lighter fragments.

Low-energy cosmic rays induce radiation with the interaction of their neutrons with boron, producing ionizing particles. Very low energy neutrons ($\ll$ 1MeV) react with the nucleus of ^{10}B, which breaks, releasing energy in the form of a ^{7}Li recoil nucleus and an alpha particle. The alpha particle and the lithium nucleus generated from the absorption of the neutron by the ^{10}B are launched in opposite directions in order to conserve momentum. They are both capable of inducing soft errors, especially in new technologies of lower voltage.

3.3 Radiation Effects

Radiation can affect electronic devices in multiple ways, as expressed in Fig. 3.3. Single event effects (SEE) are non-cumulative and are caused by single events that trigger transient upsets. Total ionizing dose (TID) and displacement damage (DD) are cumulative, which means their effects get worse over time as the system is exposed to radiation. Notice that not all radiation effects are ionizing. As will be further detailed, DD is caused by the kinetic energy of particles.

The rate at which soft errors occur in a system is called soft error rate (SER). SER is caused in semiconductor devices mainly because of three sources of radiation: alpha particles, high-energy cosmic rays, and low-energy cosmic rays [4]. An ion traveling through a silicon substrate loses energy, generating one electron-hole pair for each 3.6eV lost. The linear energy transfer (LET) of an ion defines how much it can interfere with the proper device operation. It depends not only on the mass and energy of the particle but also on the material it is traveling in (represented in units of MeV cm^2/mg).

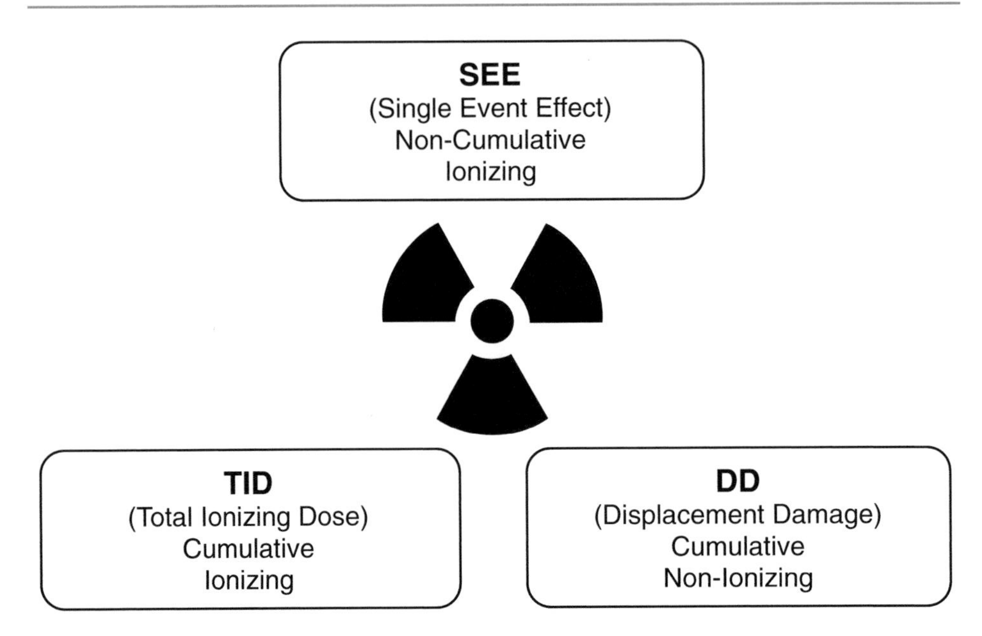

Fig. 3.3 Radiation effects in electronic devices. *Source* Author

The **faults** induced by radiation can become **errors** that might evolve into **failures**. By definition, a fault is the event itself, manifested as a bit-flip on a memory component, for example. The error, on the other hand, is the effect of the fault on the system. It can pass by unperceived or be masked by a fault tolerance mechanism. When the system misbehaves, and this is noticed by the user or propagated to another part of the system that, in its turn, shows a problematic external behavior, we say that a failure happened. Taking the example of a fault affecting the memory circuit, the definition of the events would be the following: the bit-flip on the memory data is a fault, the error is the impact it has on the data being stored, and the failure would be the malfunction of a software that could, for example, use this data as a control variable of a loop, causing the application never to finish its execution. Notice that the fault could have happened in an unused part of the hardware memory, causing no errors. Similarly, the error could have been overwritten by a store instruction shortly after happening and never turned into a failure.

3.3.1 Single Event Effects (SEE)

When radiation particles transfer enough energy into the silicon of circuits, they generate transient upsets. Upsets are manifested as bit-flips in any part of the circuit that holds data, causing errors [4, 15]. In microprocessors, bit-flips can occur in all registers and memories of the processor.

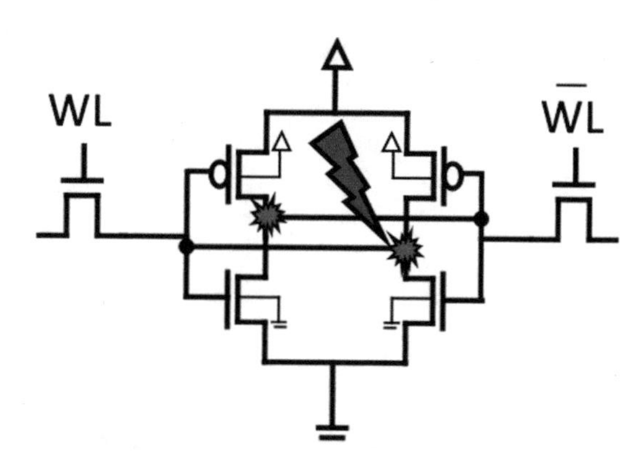

Fig. 3.4 Single event upset. *Source* Author

Integrated circuits operating in radiation environments are well known to be susceptible to errors due to particle ionization, known as soft errors. Soft errors may occur when a single ionizing particle strikes a sensitive region at the electronic device, creating a current pulse, and they have a transient effect. When this current pulse occurs in a sequential logic gate near the drain of a transistor in its off-state mode, it may flip the value stored in the memory cell, causing a Single Event Upset (SEU). When the pulse occurs in a combinational logic gate, it may propagate through the logic, causing a Single Event Transient (SET).

An SEU (Fig. 3.4) is the inversion of the value stored in a memory element, this inversion of values is commonly called bit-flip. This fault is temporary since it can be corrected with the following value written in the memory element; however, if this initial fault is propagated, it can generate an error in the execution of the application. An SEU occurs when a particle collides with a sensitive area of the memory element and deposits a sufficient minimum charge on the material to cause a bit-flip. This element can be a dynamic memory (DRAM), static memory (SRAM), or a type of flip-flop, for instance.

A SET (Fig. 3.5) is a voltage/current transient disturbance generated when an energetic particle strikes a sensible node in a combinational part of the integrated circuit. With the constant miniaturization of the size of CMOS technology, it has become clear that SET has become a significant mechanism in error rates. The scaling of the technology came along with higher operating frequencies, lower supply voltages, and smaller noise margins, making the circuit sensitivity to SET higher. Any node in the combinational circuit can be affected by a fault and generate a transient disturbance in the current/voltage with sufficient duration to be propagated along until it is captured by a memory element. However, only a few transients are captured. The chance of a SET being captured involves the probability of a functionally sensitive SET path between the node and the sequential element; the rate at which the SET loses force at each logical level that it traverses until it reaches the sequential element; and the probability of the generated transient pulse being effectively captured and stored in the sequential element. Not all SETs will be captured by sequential elements because they can be masked logically, eclectically, or by a clock window.

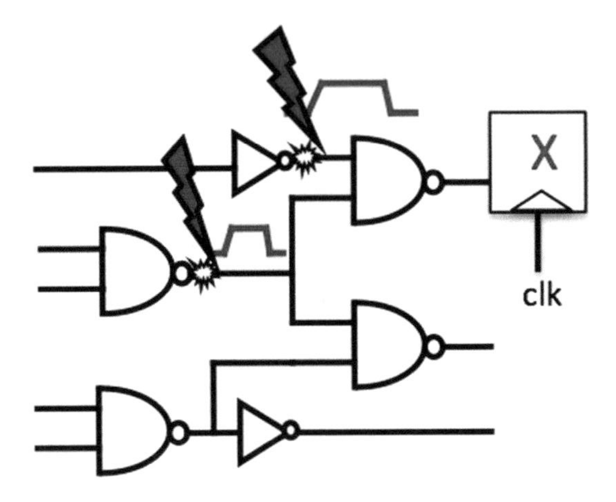

Fig. 3.5 Single event transient. *Source* Author

Table 3.1 summarizes the different SEE types and their characteristics. The characteristics exposed at Table 3.1 stand for

- **Non-Destructive:** An SEE that does not cause permanent damage to the system.
- **Destructive:** An SEE that can cause permanent damage to the system.
- **Reset Needed:** The SEE requires a full reset of the system so that it can vanish (i.e., it is a not permanent error).
- **Power Cycle Needed:** A simple reset of the system might not be enough to clean the error. Those errors are often related to physical problems affecting the operation of the transistors or logic gates.

Notice that a SEE can be both destructive and non-destructive: some of them can be destructive only in some cases, normally related to the intensity or locality of the fault. The SEE types presented in Table 3.1 are defined with more details below.

- **Single Event Upset (SEU):** As soft errors are commonly called. Those are non-permanent errors affecting one single bit of one word of data.
- **Multibit Upset (MBU):** An MBU occurs when the radiation event has enough energy to flip multiple bits on a single word. This can be especially problematic for memory circuits that use error correction codes, compromising those that cover and mask only one bit [12].
- **Multicell Upset (MCU):** It occurs when the radiation event has energy high enough to affect multiple bits on different localities. The difference between an MBU and an MCU is that the latter consists of bit-flips affecting various parts of a system (e.g., different memory words), while the former consists of multiple bit-flips in a single word [8].

Table 3.1 SEE classification and key characteristics

SEE	Meaning	Characteristics			
		Non-destructive	Destructive	Reset needed	Power cycle needed
MBU	Multibit upset	X			
MCU	Multicell upset	X			
μSEL	Micro single event latch-up	X			X
SEL	Single event latch-up	X			X
			X		
SEB	Single event burnout		X		
SEGR	Single event gate rupture		X		
SHE	Single hard error	X			X
			X		
SEFI	Single event functional interrupt	X		X	X
SET	Single event transient	X			
SEU	Single event upset	X			

- **Single Event Transient (SET):** It is considered when transient upsets occur in the combinational logic part of a circuit. Those can lead to soft errors if propagated and latched into memory elements [5].
- **Single Event Functional Interrupt (SEFI):** When a soft error occurs in a critical control circuitry (e.g., branch prediction or jump address), it can cause the processing to misbehave to the point where its execution is compromised [11]. SEFIs lead to a direct malfunction that the user easily notices (e.g., when the application ceases to respond) instead of soft memory errors that may pass unperceived.
- **Single Event Latch-Up (SEL):** SEEs can induce a latch-up by turning on CMOS parasitic bipolar transistors between the well and the substrate [6]. SELs are debilitating because a reset (powering it off and back on again) is necessary to remove them. They can also cause permanent damage.

- **Micro Single Event Latch-Up (μSEL):** This type of latch-up is related to the reduction of the transistor size and operating voltages, something to be expected of new technologies. One of the major differences between μSEL and SEL is that the latter usually occurs in the terminals of a logic gate. In contrast, the former occurs in different areas and levels of the die, provoking different effects [2]. Also, μSEL can occur under ground-level radiation [17].
- **Single Event Burnout (SEB):** When a heavy ion passes through a metal-oxide-semiconductor field-effect transistor (MOSFET) biased in the off state (blocking a high drain-source voltage), it generates transient currents that might turn on a parasitic bipolar-junction transistor inherent to this device structure. A permanent short between the source and the drain of the MOSFET is then created due to a regenerative feedback mechanism affecting the new parasitic transistor, which increases collector currents provoking a breakdown [9].
- **Single Event Gate Rupture (SEGR):** It is provoked by a dielectric breakdown of the gate oxide, caused by heavy ions. The heavy ion accumulates charge at the Si-SiO_2 interface in the gate-drain overlap region and results in electric fields in the gate oxide that cause a localized rupture. That rupture causes a permanent short between the gate and the drain of the transistor [9].
- **Single Hard Error (SHE):** A sufficiently energetic heavy ion that strikes a MOS transistor gate can locally transfer enough ionizing dose to affect its electrical parameters permanently [7]. It consists of a total ionizing dose error from a single ion that can affect SRAM memories [16].

3.3.2 Total Ionizing Dose (TID)

Apart from non-cumulative SEE, radiation can also provoke cumulative effects. This type of effect accumulates over time, altering the regular operation of the devices. They start to become a severe problem when the accumulation of faults and errors induces failures. Such is the case of total ionizing dose (TID), caused by the same physical event that can cause SEEs: the generation of electron-hole pairs. The difference between SEE and TID is that TID is accumulated over time in the device and provokes gradual and permanent changes in the device's behavior.

TID accumulates in the device when the electron-hole pairs caused by radiation effects are separated by the electric field concentrated in the transistor's gate or field oxide. The electric field prevents the electron-hole pair recombination, and because the now free electrons have high mobility, they are swept from the oxide, leaving holes behind (with low mobility). The holes get trapped in the oxide bulk and at the Si-SiO_2 interface. Those trapped holes modify the threshold voltage of the transistor by attracting electrons in the inversion channel, affecting the drain-source current. Because the number of electron-hole pairs is directly proportional to the total amount of radiation dose to which the device is subject, this effect will increase over time and change the characteristics of the transistors. The number of electron-hole pairs also depends on the dose rate, the gate-oxide electric field, and the thickness of the oxide [1].

3.3.3 Displacement Damage (DD)

Devices can also be subject to non-ionizing cumulative effects. Non-ionizing doses deposited by radiation can be a source of failures in the form of displacement damage (DD). The kinetic energy of the irradiated particles is transferred to the material and can produce atomic displacements. The rate at which this energy is passed to the material is called non-ionizing energy loss (NIEL). The DD energy deposition per unit of mass of material can be calculated by the product of the NIEL and the particle fluence (Φ_i) [10], as expressed in (3.1).

$$DD = NIEL_i \cdot \Phi_i \tag{3.1}$$

The NIEL can be calculated as (3.2) shows, where N is the number of atoms per cubic centimeters of the area being affected, T_M is the maximum transferred energy by the collision of an ion with the atoms of the material, T is the energy of the recoiling atoms, T_d is the threshold energy for atomic displacements (21eV for silicon), $\frac{d\sigma}{dT}$ is the differential cross-section for atomic displacements, and $L(T)$ stands for the Lindhard partition factor. $L(T)$ takes into account that only some of the energy of the recoil will go into producing displacements [13].

$$NIEL = N \int_{T_d}^{T_M} T \frac{d\sigma}{dT} L(T) dT \tag{3.2}$$

Displacement damage can be caused by protons, neutrons, alpha particles, and high-energy photons. As the equations induce, the displacement damage depends on the type of particle radiation, its energy, the total dose, and radiation flux. Some device characteristics also impact the possible displacement damage, such as the operating voltage, frequency, and shielding (both intrinsic and extrinsic).

3.4 Radiation-Induced Soft Errors in Zynq-7000 FPGA

FPGAs can be viewed as two-layer devices, the Design Layer and the Configuration Layer. The Design Layer of the Zynq-7000 is the PL part. The PL is composed of Configurable Logic Blocks (CLBs), specialized circuits (e.g., embedded memory blocks - BRAMs), Digital Signal Processor (DSP) blocks, an Internal Configuration Access Port (ICAP), Phase-Locked Loop (PLL) blocks, clock trees, Power-on Reset (PoR) circuitry, and others. Programmable Input/Output Blocks (IOBs) surround the PL. They are interconnected in a matrix structure by a set of programmable interconnections, creating an array of programmable logic blocks of different types. The Configuration Layer is composed of all SRAM memory cells responsible for configuring all the Design Layer, such as the CLBs, content of BRAMs, DSPs, routing structures, clock trees, PLLs, IOBs, and others. The bitstream, a group of configuration bits loaded in the configuration memory during the device power-up for defining a specific circuit previously described with a Hardware Description Language (HDL), configures the programmable blocks and interconnections.

The Zynq device is susceptible to soft errors in the PS and PL parts. Although the different kinds of memory elements in the Zynq-7000 device have similar susceptibility to bit-flips (i.e., static cross-section in the same order of magnitude), given the significantly larger amount of SRAM memory in the configuration layer of the FPGA, we expected most faults to occur in the configuration memory. Bit-flips in configuration memory have a persistent effect. They may change the circuitry functionality leading to functional failures, timeouts, and hangs. These bit-flips can only be corrected by reconfiguring the FPGA. Silent Data Corruption (SDC) failures are likely to be caused by faults affecting data memory elements or the arithmetic and logical operations. In addition, timeouts may occur when the clock tree or the reset system is affected by a soft error.

Zynq has a dual-core embedded ARM Cortex-A9 processor. In the Cortex-A9, bit-flips can occur in all registers and memories of the processor. By definition, a fault is the event itself, manifested as a bit-flip on a memory component, for example. The error, on the other hand, is the effect of the fault on the system. It can pass by unperceived or be masked by a fault tolerance mechanism. When the system misbehaves, and this is noticed by the user or propagated to another part of the system that, in its turn, shows a problematic external behavior, we say that a failure happened. Taking the example of a fault affecting the memory circuit, the definition of the events would be the following: the bit-flip on the memory data is a fault, the error is the impact it has on the data being stored, and the failure would be the malfunction of software that could, for example, use this data as a control variable of a loop, causing the application never to finish its execution. Notice that the fault could have happened in an unused part of the hardware memory, causing no errors. Similarly, the error could have been overwritten by a store instruction shortly after happening and never turned into a failure (Fig. 3.6).

3.5 Conclusion

This chapter presented the radiation effects on electronic devices. Although knowing them and their impacts is indispensable, some of them are out of the practical scope of this book. The TID and DD are related to circuit-level hardware malfunctions manifested in

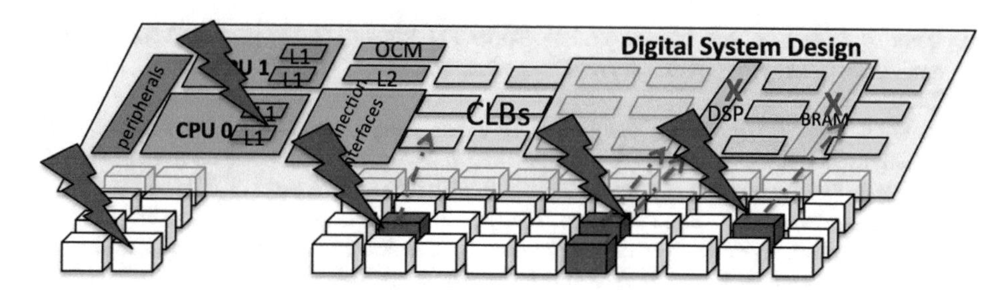

Fig. 3.6 Single event upset on FPGA. *Source* [18]

the system behavior as errors related to the intensive exposure of the device to a hazardous environment. Much like aging effects on electronic devices, those errors can hardly be dealt with by using programming methods or design strategies. Because of that, the fault tolerance and approximation techniques presented in this book will not have as their main target the effects of TID or DD. Instead, the techniques presented in this book will focus on dealing with errors caused by non-destructive SEEs such as SEUs, MBUs, SETs, and SEFIs.

References

1. C.M. Andreou, A. Paccagnella, D.M. González-Castano, F. Gómez, V. Liberali, A.V. Prokofiev, C. Calligaro, A. Javanainen, A. Virtanen, D. Nahmad, J. Georgiou, A subthreshold, low-power, RHBD reference circuit, for earth observation and communication satellites, in *2015 IEEE International Symposium on Circuits and Systems (ISCAS)* (2015), pp. 2245–2248
2. S. Azimi, L. Sterpone, Micro latch-up analysis on ultra-nanometer vlsi technologies: a new monte carlo approach, in *2017 IEEE Computer Society Annual Symposium on VLSI (ISVLSI)* (2017), pp. 338–343
3. R.C. Baumann, Soft errors in advanced semiconductor devices-part i: the three radiation sources. IEEE Trans. Device Mater. Reliab. **1**(1), 17–22 (2001)
4. R.C. Baumann, Radiation-induced soft errors in advanced semiconductor technologies. IEEE Trans. Device Mater. Reliab. **5**(3), 305–316 (2005)
5. J. Benedetto, P. Eaton, K. Avery, D. Mavis, M. Gadlage, T. Turflinger, P.E. Dodd, G. Vizkelethyd, Heavy ion-induced digital single-event transients in deep submicron processes. IEEE Trans. Nucl. Sci. **51**(6), 3480–3485 (2004)
6. G. Bruguier, J.M. Palau, Single particle-induced latchup. IEEE Trans. Nucl. Sci. **43**(2), 522–532 (1996)
7. C. Dufour, P. Garnier, T. Carriere, J. Beaucour, R. Ecoffet, M. Labrunee, Heavy ion induced single hard errors on submicronic memories (for space application). IEEE Trans. Nucl. Sci. **39**(6), 1693–1697 (1992)
8. E. Ibe, S.S. Chung, S. Wen, H. Yamaguchi, Y. Yahagi, H. Kameyama, S. Yamamoto, T. Akioka, Spreading diversity in multi-cell neutron-induced upsets with device scaling, in *IEEE Custom Integrated Circuits Conference* (2006), pp. 437–444
9. G.H. Johnson, J.M. Palau, C. Dachs, K.F. Galloway, R.D. Schrimpf, A review of the techniques used for modeling single-event effects in power MOSFETs. IEEE Trans. Nucl. Sci. **43**(2), 546–560 (1996)

10. I. Jun, M.A. Xapsos, S.R. Messenger, E.A. Burke, R.J. Walters, G.P. Summers, T. Jordan, Proton nonionizing energy loss (NIEL) for device applications. IEEE Trans. Nucl. Sci. **50**(6), 1924–1928 (2003)
11. R. Koga, S.H. Penzin, K.B. Crawford, W.R. Crain, Single event functional interrupt (SEFI) sensitivity in microcircuits, in *RADECS 97. Fourth European Conference on Radiation and its Effects on Components and Systems (Cat. No.97TH8294)* (1997), pp. 311–318
12. J. Maiz, S. Hareland, K. Zhang, P. Armstrong, Characterization of multi-bit soft error events in advanced srams, in *IEEE International Electron Devices Meeting 2003* (2003), pp. 21.4.1–21.4.4
13. S.R. Messenger, E.A. Burke, M.A. Xapsos, G.P. Summers, R.J. Walters, I. Jun, T. Jordan, NIEL for heavy ions: an analytical approach. IEEE Trans. Nucl. Sci. **500**(6), 1919–1923 (2003)
14. E. Normand, Single event upset at ground level. IEEE Trans. Nucl. Sci. **43**(6), 2742–2750 (1996)
15. C. Poivey, J.A. Barth, R. Reed, E.G. Stassinopoulos, K.A. LaBel, M. Xapsos, Implications of advanced microelectronics technologies for heavy ion single event effect (SEE) testing, in *RADECS 2001. 2001 6th European Conference on Radiation and Its Effects on Components and Systems (Cat. No.01TH8605)* (2001), pp. 328–331
16. C. Poivey, T. Carriere, J. Beaucour, T.R. Oldham, Characterization of single hard errors (SHE) in 1 M-bit SRAMs from single ion. IEEE Trans. Nucl. Sci. **41**(6), 2235–2239 (1994)
17. J. Tausch, D. Sleeter, D. Radaelli, H. Puchner, Neutron induced micro SEL events in cots SRAM devices, in *2007 IEEE Radiation Effects Data Workshop* (2007), pp. 185–188
18. Xilinx. Zynq-7000 all programmable soc overview, 2017
19. J. Ziegler, W. Lanford, The effect of sea level cosmic rays on electronic devices, in *IEEE International Solid-State Circuits Conference. Digest of Technical Papers*, vol. XXIII (1980), pp. 70–71

Methodologies for Testing and Assessing Electronic and Computing Systems

4

4.1 Summary

As discussed in previous chapters, approximate computing and fault tolerance techniques can be applied to multiple computation stacks. The same is true about radiation effects: as was explained in Chap. 3, they propagate in the system levels and occasionally cause failures and errors. The inter-stack behavior of faults and the algorithms studied in this book shall be evaluated with the appropriate use of metrics and experimental methodologies capable of emulating real-world scenarios. The book evaluates a multitude of methods and techniques, both for approximation computing and fault tolerance. Given the variety of implementations and their different implications, having one unique evaluation methodology for all of them would be impractical. For that reason, this chapter presents a number of different experimental methods. All the methodologies here presented are applied to at least one of the proposed ideas in further chapters.

The chapter is organized as follows. Section 4.2 presents the most fundamental metrics used for faults and failures. Section 4.3 proceeds to explain a methodology to inject faults on programmable hardware projects, while Sect. 4.4 presents an approach to inject faults on embedded software, targeting the register file of the processor. Similarly, Sect. 4.5 presents a methodology to inject faults on the processor register file, but this time using software simulation. A laser fault injection methodology targeting embedded processors is presented in Sect. 4.6.

All the techniques proposed in this book are implemented and tested in the same hardware. The designs are implemented in a Zynq-7000 APSoC, designed by Xilinx, represented in Fig. 4.1. The Zynq board has embedded a high-performance ARM Cortex-A9 processor with two cache levels on the processing system (PS), alongside a PL layer. The PL presents an FPGA based on the Xilinx 7-Series with approximately 27.7 Mb configuration logic bits, 4.5 Mb block RAM (BRAM), 85 K logic cells, and 220 DSP slices, with a frequency of 100

G. S. Rodrigues et al., *Approximate Computing and its Impact on Accuracy, Reliability and Fault-Tolerance*, Synthesis Lectures on Engineering, Science, and Technology,
https://doi.org/10.1007/978-3-031-15717-2_4

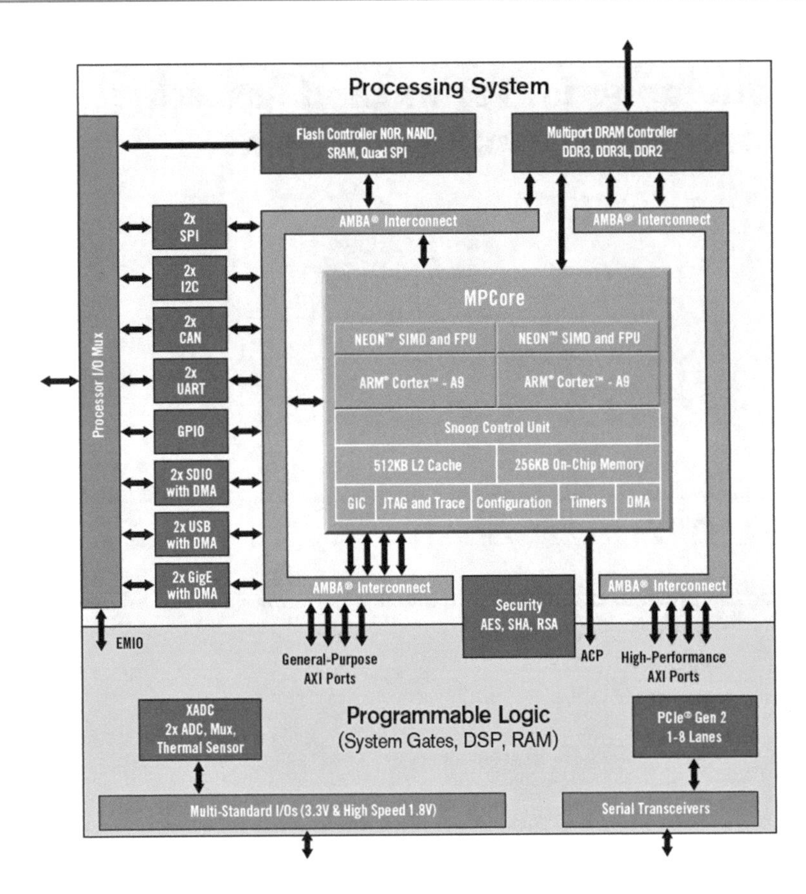

Fig. 4.1 The Zynq-7000 APSoC. *Source* [11]

MHz. The dual-core 32-bit ARM Cortex-A9 processor runs a maximum of 666 MHz and is designed with 28 nm technology. It counts with two L1 caches (data and instruction) per core with 32 KB each, and one L2 cache with 512 KB shared between both cores. A 256 KB on-chip SRAM memory (OCM) is shared between the PS and PL parts, and so is the DDR (external memory interface).

4.2 Fault Tolerance Metrics

The reliability of a system to radiation-induced transient faults can be measured in many different ways, depending on the available data and performed experiments. Some of the most used metrics for reliability and fault tolerance of safety-critical systems under radiation are the mean work to failure (MWTF) [7], the cross-section, and failure in time (FIT) [1], alongside with the already discussed soft error rate (SER). The cross-section (σ) is defined as the area of the device that is sensitive to radiation, with (4.1). A larger cross-section means that a particle that hits the device is more likely to produce a failure. Thus, a design of a smaller area (such as an approximate one) will typically present a smaller cross-section. The FIT is common as a means to express SER and is equivalent to one failure in 10^9 hours of device operation. MWTF is particularly interesting for our discussion because it presents a correlation between performance and the fault tolerance of a technique, and is presented in (4.2).

$$\sigma = \frac{\text{number of errors/failures}}{\text{fluence of particles}} \tag{4.1}$$

$$MWTF = \frac{\text{amount of work completed}}{\text{number of errors encountered}} \tag{4.2}$$

When analyzing data from simulation experiments, the error occurrence is often presented as a simple percentage. In this type of analysis, faults are injected into the system, and it is often possible to trace the types of errors and their origin. Thus, it is easy to calculate the percentage of faults that caused errors (and failures) and their types. When analyzing fault tolerance techniques, especially those implemented on embedded software, metrics like cross-section might not be the most appropriate (in fact, using this type of metric would need an adaptation, because there is no particle fluence in this type of experiment). In those cases, data might be better presented merely as the reduction of the percentage of faults capable of inducing failures.

4.3 Onboard Fault Injection Emulation on FPGA

Phenomena such as power glitches, electromagnetic interference, and ionizing radiation can cause transient effects on electronic devices. Considering storage elements, such as flip-flops and SRAM memory cells, those effects may cause bit-flips, which are the change of the storage value from a logical zero to a logical one, or vice versa. Fault injection emulation can be used as a means to assess the reliability of different designs and fault tolerance techniques. SRAM-based FPGAs have a massive amount of SRAM memory conveying the configuration information required to program the general-purpose FPGA to a specific function. On Xilinx SRAM-based FPGA, such configuration memory is organized in rows, columns, and frames. Each frame includes 101 words of 32 bits, defining the configuration for key FPGA elements such as signal routing switch boxes, multiplexers, and combinational logic truth tables.

The onboard fault injection methodology used in this book to inject faults on programmable hardware is based on the one presented by [9]. It explores the use of the Xilinx internal configuration access port (ICAP) hardware module, available on Xilinx's 7 Series FPGAs and APSoCs, to read and write at configuration memory frames. To inject a fault on a specific bit, the frame is read by the fault injection engine using ICAP, the bit content is XOR'ed, and then the frame is written back to the configuration memory. The fault injection engine is implemented on the same FPGA as the DUT but isolated from it with proper floorplanning of the design. It communicates with a host computer that coordinates the injection via a serial port. A campaign planning script running on the computer defines where a bit-flip has to be injected. Two types of fault injection are performed using this method:

- **Random Accumulated Fault Injection:** Bit-flips caused by ionizing radiation are emulated by injecting faults randomly over the area allotted to the DUT. These bit-flips are accumulated over time, as would happen if the system were under ionizing radiation. When an error is detected, the number of faults accumulated until that point is recorded, and the FPGA is re-programmed, cleaning up all previous bit-flips. This procedure is repeated until a sufficient number of errors is collected, allowing statistical analysis of the design reliability. In that type of injection, it is possible to analyze the fault tolerance of the DUT under an accumulated number of faults and how a different number of faults may impact the output of the system.
- **Exhaustive Fault Injection:** The exhaustive fault injection consists of injecting faults on every bit of the DUT configuration memory. The effect of the fault on the system output is then evaluated. This type of fault injection allows the categorization of the bits as essential and critical bits [9]. Essential bits are those that are used to program the FPGA. A critical bit is an essential bit on which a bit-flip will provoke an error in the system.

4.4 Onboard Fault Injection Emulation on Embedded Processor

In this book, the ARM processor embedded in the Zynq-7000 APSoC (Fig. 4.1) will be used as the target for the software implementation experiments. As Fig. 4.1 shows, the board has two embedded ARM Cortex-A9 processor cores. The MPCore (the unit that contains the processor cores, cache memory, OCM, and some configurations and processing units that are out of the scope of this book) communicates with external peripherals through the ARM advanced microcontroller bus architecture (AMBA). AMBA defines the communication standards and defines the AXI interface for communication with the programmable logic layer of the board.

When analyzing embedded software projects, one of the best ways to assess fault tolerance is by injecting faults in the processor register file. The register file area of the ARM processor is physically minimal. Because of that, physical fault injection experiments (e.g., heavy-ions radiation) targeted to inject specifically in the register file are often impracticable. This book proposes a methodology that uses a register file fault injector implemented at the FPGA layer of the Zynq-7000 APSoC. The injector shall access the register file without compromising the normal system execution in order to perform a trustworthy register file fault injection. This is assured by the AXI protocol and the fault injector design, which only accesses the register, which will be targeted by the fault injection at the moment. The adopted methodology follows the scheme presented in [3]. The fault injection emulation system consists of the following modules:

- **Injector Module:** Intellectual property (IP) designed in hardware description language and implemented in the FPGA layer of the Zynq board. It is responsible for performing the fault injection procedure to be detailed further.
- **Power Control:** Electrical device in charge of powering up the board in each injection cycle.
- **System Controller:** Software application running on a host computer responsible for Power Control management. It also saves the fault injection logs, which are transmitted through serial communication.

Figure 4.2 presents the experiment setup environment. The Zynq board and the power control are connected to a host computer. The host computer is responsible for controlling the system and registering experiment logs. A USB-TTL Converter is responsible for transmitting serial data containing information about the error and is connected to the DUT and the host computer.

The injector randomly injects bit-flips on the processor's register file. The affected ARM registers are the general-purpose ones, from r0 to r12, and the specific ones, which are

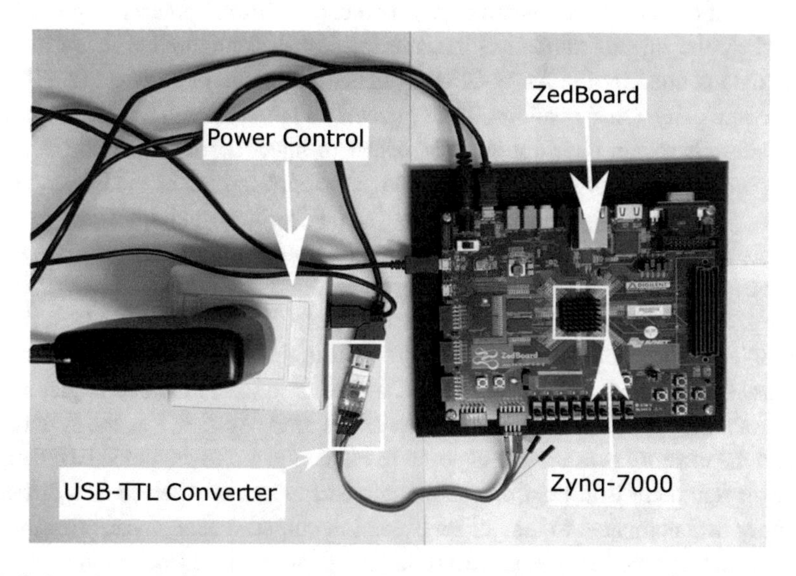

Fig. 4.2 Onboard register file fault injection setup view. *Source* Author

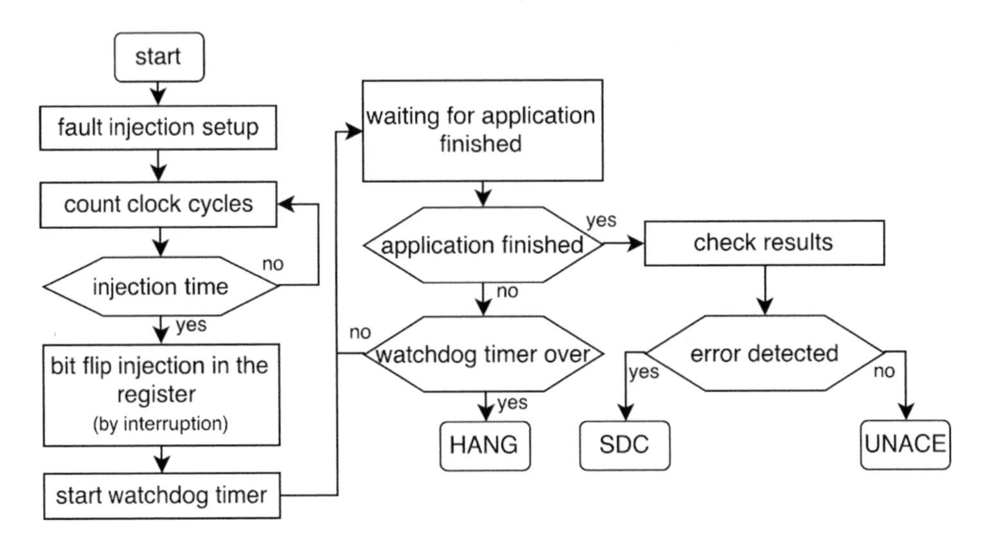

Fig. 4.3 Onboard fault injection emulation procedure flow. *Source* [3]

the stack pointer (sp), link register (lr), and program counter (pc). The faults are injected using an interrupt mechanism that locks the processor and applies an XOR mask to the target register, provoking a bit-flip. The target register and bit to be flipped are randomly defined. The injection time is also randomly determined, being a random point between the start and the finish of the software execution. It is intended to simulate real scenarios, where the fault can affect the system at any moment. Figure 4.3 presents a procedure flow performed by the injector module. First, the injector is configured with all the random injection data defined by the ARM CPU, as generating random numbers in FPGA logic has high complexity. Once configured, the injector counts clock cycles until it reaches the injection time. Next, an interruption is launched to inject the bit-flip into the processor register defined in the configuration. At the end of the application, the module compares the application results with a golden execution output (i.e., without fault injection) to check for errors. This emulation fault injection can be programmed to inject one or more faults per execution of the benchmark algorithm.

Figure 4.3 shows that there are three possible classifications to be made at the end of the fault injection regarding the impact of the fault in the system: hang, SDC, and unace. Those classifications set a standard that is to be used in the other methodologies as well. A unace means that the injected fault caused no errors to the system, i.e., the system did not crash, and the memory is as it was supposed to be (compared to the golden execution). An SDC means that there is at least one silent data corruption, manifested as a difference in the memory data compared to the golden execution output. A hang means that the system crashed, becoming wholly unresponsive or initiating an infinite execution loop.

4.5 Fault Injection Simulation

OVPSim [4] is a full-system simulator used to simulate the execution of code in the target hardware. It uses a just-in-time binary translation, achieving high simulation speeds. That makes OVPSim an instruction-accurate simulator, providing the possibility to analyze the execution of each individual instruction, but not real execution times (e.g., clock cycles). OVPSim provides public APIs which allow full control of the target simulation and the implementation of the fault injection module. The OVPSim was chosen instead of other popular cycle-accurate simulators (like gem5) because it provides a more reliable processor model. Gem5 targets microarchitecture exploration, which incurs substantial simulation overheads due to the number of modeled aspects. OVPSim also provides better APIs for simulation development than other simulators. Moreover, it has an active development and support team.

The application executions are simulated on a single-core ARM Cortex-A9 model. This processor was selected due to its presence on COTS devices that are used for safety-critical applications, such as the Xilinx Zync-7000 APSoC, which is extensively used in this book. The model used to simulate the ARM Cortex-A9 was the one developed to be specially used at OVPSim, with ARMv71 architecture. This model is extensively used and validated by embedded software developers, which use the OVPSim simulator to test their projects.

The OVPSim-FIM (OVPSim Fault Injection Module), developed by [8], was slightly modified and configured to be used for fault injection and error evaluation. The fault injection simulation will only inject faults in the processor register file. A fault is modeled as a bit-flip, to be injected into a register in a certain instruction count (ICOUNT). The ICOUNT holds the number of instructions that were executed so far by OVPSim. Because OVPSim is instruction-accurate and works with just-in-time binary translation, the best way to define the injection moment is with an ICOUNT, which represents an execution point in time.

OVPSim runs the simulation of the benchmark applications execution on the processor model, finishing the simulation after it reaches a predefined point that is hard-coded in the application code. OVPSim-FIM injects a single fault per simulation execution. All faults are randomly generated; that is, they are random bit-flips, in a random register, in a random ICOUNT. OVPSim-FIM counts with a fault monitor function, which checks the system behavior, dynamically detecting some types of errors, such as hangs (defined at Table 4.1). The results are generated comparing the executions under the effects of fault injections with an error-free run (golden execution) of the system.

The proposed simulation fault injection methodology is divided into five phases according to fault injection activities and result gathering:

1. **Golden Phase**: In this phase, the application is executed with no fault injections. It is the pure execution, to be taken as a reference run of the application (i.e., the golden run). The applications are compiled using the proper cross-compiler, then executed in the OVPSim simulator. The execution output is saved in a report called "golden report". This report

Table 4.1 Error type classifications for simulated fault injections using OVPSim-FIM

Group	Error definiton
Hang	Causes the application to be stuck in a certain point
SDC	Difference in the final memory from the golden phase execution and the fault-injected test executions
Unace	Injected fault caused no error
Exception	Error captured by the operating system

contains vital information, such as the first and final ICOUNT, the memory dump of the application after the golden execution, and the final values of the registers. The first and final ICOUNT define the possible fault injection window and are set by two instructions that are added to the application code. Those instructions are added at the beginning and the end of the code and don't count as possible injection points. They are used solely for fault modeling purposes and do not affect the final application solution or execution.

2. **Fault Creation Phase**: This phase is where faults are created using the information gathered in the previous phase. A single bit-flip in the logical registers is provoked by an XNOR operation with a mask. The ICOUNT data of the golden phase is used to determine the possible fault injection execution times. OVPSim-FIM calculates a random injection execution time (i.e., ICOUNT), limited between the first and the final ICOUNT of the golden phase execution. The fault location (that is, the register where the fault is to be injected) is also randomly defined. A fault list is then created. Each fault is represented by a structure containing an ICOUNT, a register, and a mask.
3. **Execution Phase**: At the execution phase, the fault injection module monitors the current ICOUNT while the application is executed in OVPSim, and injects the fault at the fault's ICOUNT, defined at the fault creation phase. The targeted register is then accessed, the mask pattern applied, the resultant value written in the register, and the application execution resumes. OVPSim-FIM system keeps track of the current ICOUNT even after the fault injection, to deal with possible hangs. If an application executes at least 50% more instructions than the golden execution total ICOUNT, it is considered to have a hang error, and its execution is halted (otherwise, the simulation could execute indefinitely). The OVPSim-FIM also watches for possible OS exceptions, such as segmentation faults. When this type of error happens, it halts the simulation and reports the problem.
4. **Error Detection Phase**: In this phase, a comparison between the golden phase execution (golden execution, fault-free) and the execution phase (faulty execution) executions is made. Errors are classified in four different groups, defined in Table 4.1.
5. **Final Phase**: At this phase, the results from all fault injections and errors found are grouped in a single report, which is the one used to generate the final data.

The OVPsim-FIM test reports present over sixteen types of errors. For practical reasons, those types of errors are re-classified into four different groups: hang, SDC, unace, and exceptions. Those groups are defined in Table 4.1. As will be discussed in Chap. 5, the exception error group can be further categorized to provide a better view of the types of errors.

4.6 Laser Fault Injection

Laser testing is commonly used as an in-lab tool for injecting transient localized perturbations into a device under test by photoelectric stimulation, especially for SEE investigations [2], security evaluation [10], and more generally to evaluate the fault tolerance of an application [5].

The fault injection experiments were performed on the two-photon absorption (TPA) microscope of IES laser facilities at the University of Montpellier. The TPA method was preferred to the more classical single-photon approach because it provided a better reproducibility of the fault occurrences in the 28 nm DUT (dual-core 32-bit ARM Cortex-A9 processor embedded at the Zynq-7000 APSoC) with a thick substrate (700 μm). The laser wavelength is 1064 nm, with a pulse duration of 30ps, and the beam was focused through the backside of the DUT by a 100× lens. The DUT was scanned under the static beam using motorized stages.

The laser pulse energy is set at 250pJ. This value was found in the literature [6] to induce between 0 and 3 bit-flips per pulse in the region of interest of the DUT, depending on the laser position. This energy is slightly above the energy threshold for a single bit-flip. Laser pulses are triggered with a frequency of 10–20 Hz (depending on the benchmark under execution) without any synchronization with the DUT clock, while the target application is executed in a loop by the DUT. Making use of this asynchronous approach, the methodology assures randomness concerning the arrival time of the laser pulse and the current application execution cycle. This allows the statistical coverage of any vulnerability time window without the need for unreasonable experiment time. Figure 3 presents a microphotograph of the processor with the fault injection regions highlighted. Two Regions Of Interest (ROI) are defined: one to cover the L1 data caches of both processors (215 × 780 μm^2) and the other that covers the OCM (two areas of 820 × 440 μm^2 each). Both the ROIs were scanned repeatedly for each benchmark execution with a step of 2 μm and 3 μm, respectively, for the X- and Y-axes. Considering the constant pulse triggering rate, the maximum scanning speed along the x-axis was adjusted to have at least one pulse every μm along x. Due to the acceleration and deceleration phases at the extremities of each scan line, this approach leads to smaller steps along the x-axis between consecutive pulses near the edges of the ROI. This approach was preferred to a strict step in order to maintain a constant laser pulse rate, and thus an accurate control of the number of pulses per application execution cycle. Indeed, in this book, we are more interested in time-related statistics than in the accurate

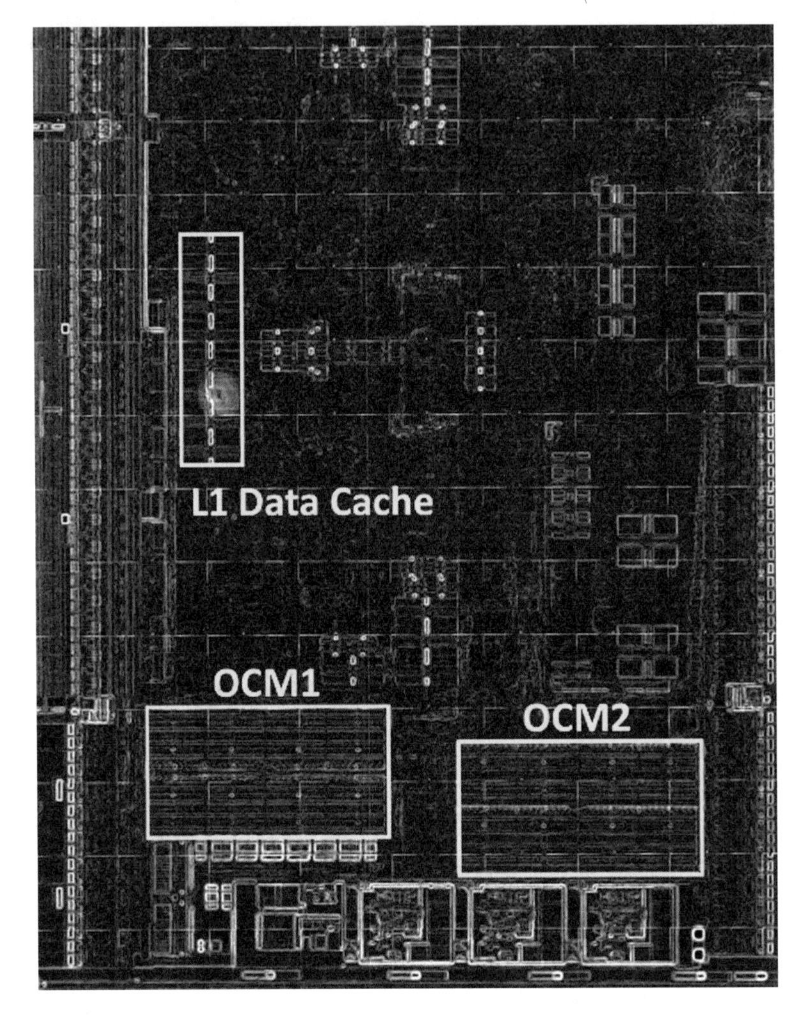

Fig. 4.4 Infrared microphotograph of the DUT core under test, showing the scanned areas (L1 Data Cache and OCM). *Source* Author

spatial localization of the occurrence of the fault. Notice that the DUT has only one OCM of 256KB, as explained before, but this memory is scattered in the two areas that were defined as one single ROI and presented in Fig. 4.4.

The experiment setup is presented in Fig. 4.5. It consists of the DUT, a host computer, and the laser equipment. The host computer is responsible for controlling the laser beam and listening to messages from the DUT. The DUT periodically sends messages to the host computer, to report an error or to confirm it is alive. Error messages are reported when there is a difference between an execution output and the golden output. The golden output is the result of a fault-free execution at the beginning of the experiment, called golden execution.

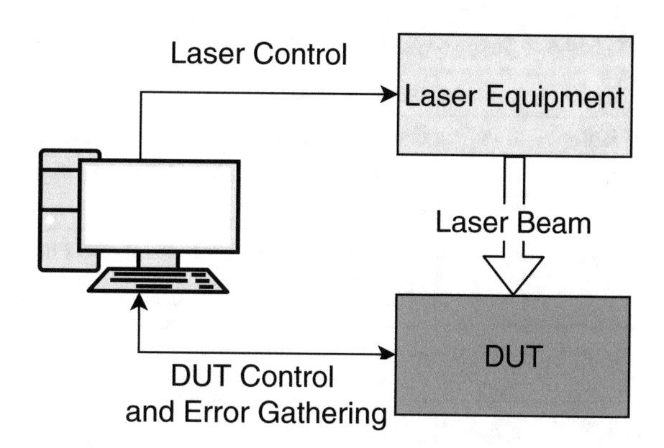

Fig. 4.5 Laser experiment setup. *Source* Author

The alive message is essential because some faults will cause the DUT to be irresponsible or hang, needing a reset. A reset consists of re-programming and configuring the DUT and is performed when a timeout occurs, while the host computer waits for an alive message from the DUT. This timeout is set to about three minutes but may vary for different experiments with different response times. During a reset, the DUT warns the host computer so that the laser beam is deactivated. It prevents any errors during the system initialization and golden execution. The laser beam is then reactivated after the host computer receives an alive message from the DUT.

The communication between the host computer and the DUT is rather complex and is highly susceptible to errors because it happens during the fault injection. To avoid errors that are not interesting to our experiment and would make it less efficient, we developed a strategy to reduce this communication to a minimum necessary. During the benchmark execution, the algorithm runs N times, filling in an output vector, which will be then compared with the result from the fault-free execution (golden value). This way, the DUT only has to send messages to the host computer every N runs (and when an error is detected). The value of N may vary for different benchmarks, according to their execution times (those details will be presented at Chap. 7). After each algorithm execution, the output vector is compared with the golden value to check for its correctness. When the output value and the golden value are different, the DUT sends to the host computer a message containing the details of the error (position on the output vector and the value of the incorrect output). The host computer receives the error messages and saves them into a log to be further analyzed. The errors are classified into three different types: hang, SDC, and multi-SDC. They are defined in Table 4.2. It is important to notice, however, that depending on the objective of the experiment and the system being tested, the studied error types might be different. For instance, a given fault tolerance technique may consider for its coverage analysis Multi-SDCs and SDCs as of equal importance, and present the numbers of both of them simply as SDCs. Table 4.2 merely presents the error types that this fault injection methodology can differentiate.

Table 4.2 Error type classifications for laser fault injections

Type	Error definiton
Hang	Causes the application to be stuck at a certain point
SDC	Output difference between the golden execution and the one exposed to laser fault injection
Multi-SDC	Multiple SDC occurrences in the same run, e.g.: multiple positions of the output vector corrupted

4.7 Conclusion

This chapter presented an overwhelming amount of testing methodologies and evaluation metrics. One might be tempted to find a favorite among them and defend its use as the ultimate testing methodology. However, this is not the ideal path. Each one of those methodologies has cases in which they are more or less appropriate.

Laser fault injection, for example, is ideal when we want to study the behavior of the system when faults affect a limited part of it. Or else to evaluate the dependability of a given memory area that is hardened by design. Such tests would be much harder to do using fault injection simulation because a simulation cannot fully represent the behavior of silicon devices and how they react to radiation (they can go so far, and even so, at very high computational costs). It is, therefore, crucial to be aware of the differences between the methodologies: using the wrong one might give misleading results.

References

1. R.C. Baumann, Radiation-induced soft errors in advanced semiconductor technologies. IEEE Trans. Device Mater. Reliab. **5**(3), 305–316 (2005). (Sept)
2. S.P. Buchner, F. Miller, V. Pouget, D.P. McMorrow, Pulsed-laser testing for single-event effects investigations. IEEE Trans. Nucl. Sci. **60**(3), 1852–1875 (2013). (June)
3. Á.B. de Oliveira, L.A. Tambara, F.L. Kastensmidt, *Exploring Performance Overhead Versus Soft Error Detection in Lockstep Dual-Core ARM Cortex-A9 Processor Embedded into Xilinx Zynq APSoC* (Springer International Publishing, Cham, 2017), pp. 189–201
4. Imperas. Open virtual platforms (ovp), 2017
5. V. Pouget, A. Douin, G. Foucard, P. Peronnard, D. Lewis, P. Fouillat, R. Velazco, Dynamic testing of an DRAM-based FPGA by time-resolved laser fault injection, in *2008 14th IEEE International On-Line Testing Symposium* (2008), pp. 295–301
6. V. Pouget, S. Jonathas, R. Job, J.R. Vaillé, F. Wrobel, F. Saigné, Structural pattern extraction from asynchronous two-photon laser fault injection using spectral analysis. Microelectron. Reliab. **76–77**(Supplement C), 650–654 (2017)
7. G.A. Reis, J. Chang, N. Vachharajani, S.S. Mukherjee, R. Rangan, D.I. August, Design and evaluation of hybrid fault-detection systems, in *32nd International Symposium on Computer Architecture (ISCA'05)* (2005), pp. 148–159

8. F. Rosa, F. Kastensmidt, R. Reis, L. Ost, A fast and scalable fault injection framework to evaluate multi/many-core soft error reliability, in *2015 IEEE International Symposium on Defect and Fault Tolerance in VLSI and Nanotechnology Systems (DFTS)* (2015), pp. 211–214
9. J. Tonfat, L. Tambara, A. Santos, F. Kastensmidt, Method to analyze the susceptibility of HLD designs in SRAM-based FPGAs under soft errors, in *Applied Reconfigurable Computing*, ed. by V. Bonato, C. Bouganis, M. Gorgon. (Springer International Publishing, Cham, 2016), pp. 132–143
10. E. Trichina, R. Korkikyan, Multi fault laser attacks on protected CRT-RAS, in *2010 Workshop on Fault Diagnosis and Tolerance in Cryptography* (2010), pp. 75–86
11. Xilinx. Zynq-7000 all programmable soc overview, 2017

Part III
Fault Tolerance and Approximation in Practice

5 Embedded Systems Fault Tolerance

5.1 Summary

Faults caused by radiation on electronic devices can become errors that must be treated before evolving into failures. The more usual way of doing it on complex systems is with fault tolerance techniques implemented in programmable hardware or embedded software [1]. Fault-tolerant and radiation-hardened devices are usually expensive. Therefore, the industry tends to turn to in-house developed fault tolerance techniques. Those techniques shall be able to detect errors and masking (i.e., correcting) them when possible.

The chapter starts by discussing how fault tolerance is implemented to protect safety-critical systems in Sect. 5.2. Most of those fault tolerance techniques are already extensively discussed in the literature. Therefore, this book will exemplify how to apply fault tolerance techniques and evaluate their performance in the case of embedded parallel software running on multicore systems. With that in mind, Sect. 5.3 presents a practical analysis of the reliability of parallel multicore systems. Section 5.4 expands this analysis by evaluating parallel fault tolerance techniques. Finally, Sect. 5.5 will finish with some conclusions and an introduction to the ideas to be explored in the following chapters.

5.2 Fault Tolerance

Figure 5.1 classifies fault tolerance techniques in three major groups concerning their tolerance capability. A fault tolerance technique must be able to detect errors. What it does with this information (i.e., an error has been detected), however, may vary. As will be further exemplified in this book, for some systems, fault detection is enough. Nevertheless, safety-critical systems often call for error masking or correction. The difference between error masking and correction is that masking an error consists of keeping the system safe

G. S. Rodrigues et al., *Approximate Computing and its Impact on Accuracy, Reliability and Fault-Tolerance*, Synthesis Lectures on Engineering, Science, and Technology,
https://doi.org/10.1007/978-3-031-15717-2_5

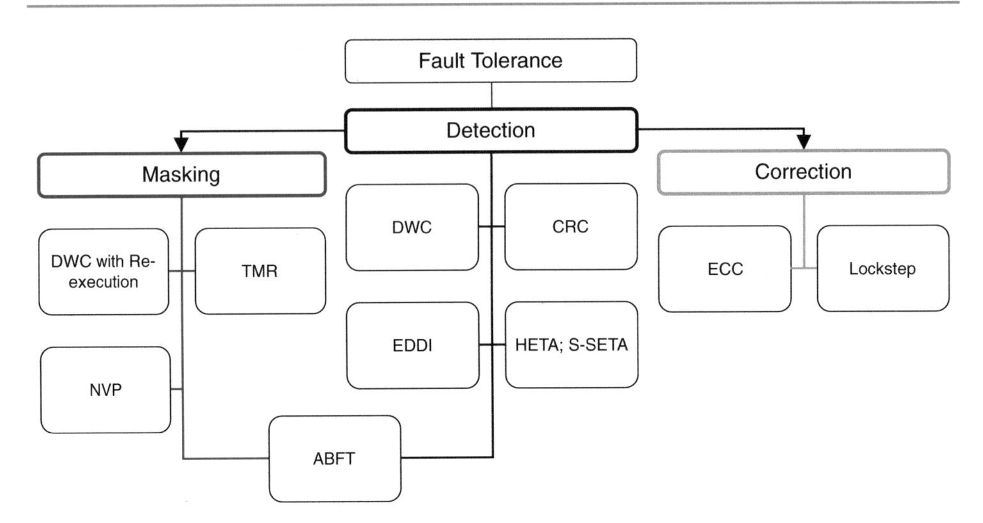

Fig. 5.1 Fault tolerance techniques classification. *Source* Author

and hiding the error from the end-user (or the rest of a more complex system). An excellent example of this type of fault tolerance technique is triple modular redundancy (TMR) [20], which avoids the use of an erroneous data value, by majority voting the correct one. Correcting an error is a much harder and complex task, and from a system point of view, the impact would be the same as masking it. As an example, the lockstep technique [7] finds an error and rolls all the system execution back to a safe-state before the error happened, and then resumes the system execution with the hopes that the error has forever vanished.

The literature presents an enormous set of techniques implemented in software to protect applications against hardware errors. Those are called software-implemented hardware fault tolerance (SIHFT) techniques [10], and achieve protection with function redundancy and variables replication. An example of this type of technique is error detection by duplicated instructions (EDDI) [17]. EDDI detects faults by comparing two different executions of the program, mapping all numbers on the original computation to new values, and applying transformations to the program so that it can be backward comparable with the original calculation. Techniques like HETA [2] and S-SETA [5] detect control-flow faults and put the system in a fail-safe state. The CFT-tool [4] is capable of combining these techniques to detect both SDCs and functional interruption (FI) errors. CFT-tool inserts the fault tolerance methods directly on the Assembly level code of the program to be protected (after the compilation). Nevertheless, it can present some limitations for complex applications and those which are supposed to run on top of operational systems. Techniques called application-based fault tolerance (ABFT) encode the used data, profiting from unique application characteristics [13]. ABFT shall be specifically designed for the application under protection. Therefore, it is not scalable to a high range of applications and tends to be costly in design. Both SIHFT and ABFT come with the cost of execution time overhead.

Redundancy-based methods such as TMR and duplication with comparison (DWC) are employed in a multitude of systems, both to provide error detection and masking. TMR can be implemented both to protect a hardware module [19] or software code [18]. It consists of triplicating the hardware (or software code) and voting the output of the redundancies: at least two over the three output replicas have to be equal and considered (i.e., voted) as the correct output. When TMR is used on hardware, it mainly implies an area overhead. When it is applied to software, it mostly provokes execution time overhead. On the other hand, DWC techniques are only capable of detecting errors, but the detection signal can provoke a re-execution of the component in order to provide error correction (i.e., in the case of transient fault, the re-execution is supposed to restore a correct state). This way, an error can be detected and mitigated before becoming a failure. DWC techniques have an overhead of two times the execution time of the original application for pure redundancy and three times when applying re-execution for error correction. N-version programming (NVP) [3] is a programming strategy that consists of developing a number of different (but equivalent) algorithms from the same specification. With this method, designers hope to achieve fault tolerance via code redundancy (voting the results from each one), expecting that two different programmers independently generating code would not produce software that is susceptible to the same errors.

Cyclic redundancy check (CRC) [14] is commonly used on network and storage systems to detect errors affecting the stored data. This error checking method is broadly used on network systems because it is easy to implement on hardware and perfect to detect burst errors, as well as those caused by noises in the transmission. A multitude of CRC designs is proposed in the literature, but it consists of check values based on the calculations of polynomials, which shall be re-calculated to verify if the check value remains the same. If it is not, there is an error in the data. CRC can be used as a first step for error correction. Error correction codes (ECC) [8] is also presented in the literature in various forms. Hamming ECC, for instance, is extensively used to protect memories (e.g., DRAM, and NAND-based flash) against errors. This method provides the correction of one error and the detection of up to two errors (with no correction possible in this case).

In [9], an approach based on task level migration is proposed as fault tolerance for aerospace FreeRTOS applications on multiprocessor systems. The technique consists of migrating tasks from a faulty processor to a fault-free one. The detection of the error is done by middleware blocks assumed to be fault-tolerant. The problem with that approach is that a high amount of assumptions must be taken beforehand by the programmer. Decisions (like which tasks will be migrated to which processor nodes) are made in the programming phase. Unfortunately, a programmer can never safely predict which processing nodes will fail.

The authors in [12] proposed a Python-based programming model (PyDac) to improve the reliability of heterogeneous many-core systems. They evaluate this proposal in an architecture composed of six ARMv2, high throughput soft cores, and one embedded PowerPC, which is optimized for single-thread performance. Running on a Linux kernel in the embedded PowerPC, PyDac decomposes the application in redundant parallel subtasks scheduled on the soft cores with Python virtual machines. The system dynamically checks the results of the subtasks to ensure the resilience of the six soft cores. However, the authors assume that the main PowerPC processor is fault-free, which is not a realistic assumption, and did not implement any technique to protect it against soft errors.

In most of the cited works, recovery approaches are proposed to deal with the error. Nevertheless, a plethora of safety-critical applications may not need recovery. As stated in [6], real-time systems have to deal with *data freshness* requirements, which defines the time interval on which data is considered valid. For instance, an automatic navigation system may have an error during its execution, but because its data freshness has a minuscule time interval, the error will soon disappear as the algorithm keeps its execution generating a new value. Because of that, an error correction procedure is not always necessary. However, the system shall be aware of the error to put itself in a fail-safe mode. Indeed, in some cases, it is better to warn the user about the error and let him decide how to handle it. Such is the case of some errors that might affect an airplane system, for example. Trying to correct an error in an airplane can cause the system to overwrite the pilot's demands and cause a catastrophe. In those cases, it is often better to warn the pilot that certain data is not to be trusted or alert for a malfunction and let him deal with the situation in the most suitable manner. This type of situation calls for an error detection system (without the masking/correction capability). In those cases, the values of the redundant re-computations are only used for comparison and error checking. That is where a designer may profit from approximate computing, as will be proposed in Chap. 6 (Sect. 6.5).

5.3 Reliability of Parallel Embedded Software on Multicore Processors

Safety-critical systems manage the execution of many resource-sharing applications. Especially for avionics applications, the designs require the Radio Technical Commission for Aeronautics (RTCA) certification to operate in most countries. For software, the DO-178B/C certificate is desired, while for hardware the DO-254 certificate is needed [11]. This certificate's exigencies are highly conflicting with memory sharing software systems, imposing a large number of limitations and safety measures to avoid catastrophic errors.

There are reasons to limit safety-critical applications for single-core architectures (alongside other hardware limitations). Performance is not the main concern for that area of computer engineering. Nevertheless, the tendency for the microprocessor industry is to turn to multicore to achieve better performance. The developers tend to focus on the satisfaction of their larger market shares. As safety-critical systems are not the most abundant consumers of the industry, they may have to adapt to make use of what the industry has to offer, which is hardware designed with a focus on performance, not on fault tolerance.

The reliability of multicore systems is a major concern for safety-critical applications. Multicore processors tend to be more susceptible to SEUs because of their high level of miniaturization and the tendency to have a large number of memory cells. On the other hand, having access to multiple processing cores opens the possibility to achieve fault tolerance via execution redundancy [16]. Given the fact that the industry has turned to multicore processors and parallelism as the *modus operandi* to provide a better performance, it is imperative to understand the implications of parallelism on the reliability of the system. This section will discuss the implications of using parallelism on safety-critical systems, and present experimental data acquired by fault injection simulation to evaluate the reliability of those systems. We first focus on the embedded multicore systems reliability study as a whole, while the second time we will present a study on the effects of using parallelism applied to the fault tolerance techniques themselves (i.e., not only in the algorithm that is being protected).

Having access to multiple processing creates an excellent environment to achieve fault tolerance via redundant execution. A myriad of application program interfaces (APIs) for parallelism is available. One of the most used APIs for that purpose is OpenMP (Open Multi-Processing), which supports multi-platform shared memory multiprocessing programming in Fortran and C/C++. OpenMP is supported by most platforms, processor architectures, and operating systems. Another highly used interface is the POSIX Threads API, usually referred to as Pthreads. Pthreads allows a program to create and manage multiple flows of execution, i.e., threads. It is available on most Unix-based operating systems, such as Linux and MAC OS. Those APIs vary in terms of abstraction level and implementation complexity. Consequently, the resultant application will differ when developed using different APIs, even when using the same parallelization strategy. When applying those applications to safety-critical systems, it is imperative to know their critical failure points and susceptibility to errors.

Bare metal applications are suitable for small systems, offering a significant degree of control during their development. However, applications' complexity increases quickly, reducing the overall system reliability. Additionally, bare metal applications developed for specific projects are more prone to development errors. In contrast, commercial operating systems (OS) are well known and tested by a large community. Thus, most of the development bugs are assumed to be fixed. To guarantee the safe management of resources, a commercial OS such as Linux is attractive. Developing a specific OS for radiation-hardened or safety-critical systems is costly. On the other hand, executing bare metal applications on a complex system could induce a waste of resources that would be better managed by an OS. Using an operating system is also a significant concern because the operating system itself may be affected by faults. The Linux OS susceptibility to soft errors has been studied by [15]. The fault tolerance of the μC-Linux embedded OS, commonly deployed in real-time applications such as the automobile industry, is also well-documented [22].

On top of that, when making use of a parallel API such as OpenMP and Pthreads, an operating system such as Linux is mandatory. Therefore, it is not a negligible part of the system, impacting directly on the system fault tolerance, and needs to be studied. This part of the book aims at investigating the parallelization paradigm effects on application reliability and their combined impact when deployed alongside a complex OS such as Linux. Additionally, traditional fault tolerance techniques' applicability is explored in each of those system configurations. For this purpose, OpenMP and Pthreads applications are evaluated with fault tolerance techniques at the software level. Additionally, the usage of redundancy TMR and DWC with re-execution, here named conditional double modular redundancy (CDMR), is investigated at the software level in single- and dual-core processors under fault injection simulation.

Fault injection simulation experiments are performed with the OVPSim, as presented in Sect. 4.5. The faults are injected into registers in single- and dual-core versions of an ARM Cortex-A9 processor during the execution of the benchmark applications. The types of errors are the same presented in Table 4.1. The errors from the exceptions group are divided into segmentation faults and unidentified errors. This categorization was made because it was clear that segmentation fault error was the most recurrent type of exception error, which makes it important. Exceptions that are not categorized as segmentation faults will be called unidentified errors. Bare Metal applications present no exception type errors, because of the absence of an operating system to catch those. The bare metal sequential algorithm versions are only tested on single-core execution, while the Linux applications are evaluated both executing on dual-core and single-core. The impact of Linux is studied by comparing the results of fault injections between the same algorithms running bare metal applications and on top of the Linux OS. As the algorithms are the same, and so are the simulation and fault injections, it is expected that all the difference between the results is provoked because of the usage of Linux OS. To provide statically relevant results, the fault injection simulation experiments were executed up to the point where the error distribution ceased to vary. Therefore, the number of experiments performed shall be enough to provide reliable results.

Three benchmark applications were tested: Bit Count, Matrix Multiplication, and Vector Sum. Bit Count is an ordinary bit count verification that counts how many bits are set in a given word. Matrix Multiplication is a simple matrix multiplication operation, and the Vector Sum is the sum of two vectors. Those are simple codes focused on pure calculations and processing, making no use of any special programming strategy.

The first analysis regards the sequential versions of the benchmarks. Two versions of each sequential algorithm are presented. The first one is a bare metal application implementation, which is executed on top of no operating system. Because there is no operating system, there is no process dedicated to the management of the two available ARM cores. Bare metal applications presented in this chapter were executed using only one processor core. The same cannot be said about software that is executed on top of Linux. When running on top of an operating system, an application is subjected to this system's scheduling and resource management algorithms. Considering that a comparison between a dual-core and a single-core execution would be unfair, the results from a single-core version of the Linux execution are presented (i.e., single-core ARM Cortex-A9). In this first analysis, a preliminary idea of the impact of using an operating system on the application fault tolerance is studied. Table 5.1 presents the results of those injections.

The OpenMP and Pthreads parallel versions of each benchmark application are also executed and compared. Those versions of the application are always executed on top of Linux OS (on both cores of the dual-core processor). The resource management is entirely done by the Linux OS. Sequential versions of the algorithms were also executed, but this time on the dual-core processor model. The results from this dual-core sequential execution are used as a reference to evaluate the impact of using the parallelization APIs on the system's fault tolerance. Table 5.2 presents the results of those simulations.

From Table 5.1, it is clear that the impact of Linux OS is different depending on the workload. Still, the usage of Linux OS on a single-core processor making no use of parallelization APIs seems to have little influence on the overall fault tolerance. Bare metal applications present no exceptions, because of the absence of an operating system to catch them. However, this information alone is not enough to say that they are less susceptible to errors, for exceptions presented in Linux may manifest themselves as other kinds of errors in bare metal applications. Supporting that theory, the percentage of hangs is lower on Linux. This is probably because the operating system is catching those errors and handling them as exceptions instead of letting them turn into hangs.

It is evident by the results of Table 5.2 that the usage of parallel code increases the occurrence of SDC type errors, the Pthreads versions being more susceptible than the OpenMP ones. Parallel applications are also more susceptible to Segmentation Fault errors. That increase in the percentage of errors in parallel applications justifies the development of special fault tolerance techniques to handle those different applications. The speedups presented were gathered from the real execution on the Zynq-7000 board (i.e., not simulation) and are calculated in relation to the sequential version of the given benchmark application. It is noticeable that the Pthreads applications achieved superlinear performance gains. Although

Table 5.1 Fault injection results in percentage of errors, running sequential applications on single-core ARM Cortex-A9

		Errors [%]				
		Unace	SDC	Hang	Exceptions	
Application	Version				Seg. fault	Unidentified
Bit Count	Bare metal	45.6	40.9	13.5	–	–
	Linux	40.4	42.1	0.2	17.0	0.3
Matrix Mult.	Bare metal	58.3	31.3	10.4	–	–
	Linux	34.3	46.8	0.4	17.8	0.7
Vector Sum	Bare metal	56.4	32.4	11.2	–	–
	Linux	55.7	25.7	0.4	17.4	0.8

Table 5.2 Fault injection results in percentage of errors, running sequential and parallel applications on dual-core ARM Cortex-A9

		Errors [%]					Speedup
		Unace	SDC	Hang	Exceptions		
App.	Version				Seg. fault	Unidentified	
Bit count	Sequential	51.2	30.9	5.9	10.3	1.7	1.00
	OpenMP	47.8	28.6	3.7	19.7	0.2	1.96
	Pthreads	37.5	46.4	0.6	15.2	0.3	1.09
Matrix mult.	Sequential	45.4	37.0	4.8	12.3	0.5	1.00
	OpenMP	33.4	42.6	4.7	19.1	0.2	2.02
	Pthreads	32.6	45.6	1.1	20.4	0.3	2.34
Vector sum	Sequential	43.6	40.2	4.9	10.9	0.5	1.00
	OpenMP	29.7	47.4	4.7	18.0	0.3	1.81
	Pthreads	31.6	47.5	1.4	19.1	0.4	2.56

the unaces decrease using Pthreads compared to sequential, the speedup factor helps to significantly reduce the exposition time of the application under soft errors.

Comparing the executions on Linux and bare metal, the Matrix Multiplication running on Linux has fewer unaces than their bare metal counterparts, which means a higher number of errors. This is also true for the other two benchmarks. Other related works have seen similar behavior, such as [21].

The results from the execution on top of Linux OS using a single-core processor and on a dual-core processor for a sequential application present exciting data. The dual-core execution was less susceptible to errors than the single-core execution, except for the Vector Sum algorithm. It happens because faults are injected in both processors, so the probability for a fault to hit a vital instruction is higher on single-core execution when executing sequential algorithms. The comparison between the executions of the sequential version of the applications and the parallel ones in the dual-core processor, running on top of Linux OS, clearly shows that parallel executions are more susceptible to errors than their sequential counterparts. Some of the benchmarks presented almost double the segmentation fault exceptions on the parallel version.

Results show that parallel applications are more susceptible to errors than their sequential counterparts, presenting higher numbers of SDC and Segmentation Fault errors (except for Bit Count OpenMP version). Remember that even when executing sequential algorithms, with Linux in dual-core processors, both cores may be in use (because of the OS scheduler). This is conflicting with some of the radiation results from the experiment from [23], where using parallelism implied a smaller cross-section when compared to the use of a single-core processor. Still, their work differs from ours in many ways, notably in the sense that our system under evaluation is more complex (with an OS and dual-core ARM processor).

After the results from Tables 5.1 and 5.2, it becomes clear that parallel applications executing on multicore architectures need to be protected at least as much as sequential ones. The fact that parallel applications behave differently from their sequential counterparts when exposed to faults raises the question of whether fault tolerance techniques classically used to protect sequential algorithms would have the same efficiency when applied to parallel applications. In an effort to answer that question, we evaluate two fault tolerance techniques:

- **Triple Modular Redundancy (TMR)**: The TMR technique applied to embedded software consists of running the part of the code to be protected three times. Each of those executions saves the computed data in different memory spaces. After the executions, the three values are compared and checked for inconsistencies. It is essentially a software adaptation of the TMR technique used for programmable logic devices. If no difference between the values is found, then no error is to be found by the technique. If one of the values differs from the other two, then it is considered erroneous, and it is masked (by considering the other values as the correct output and discarding the different one). TMR may cause a penalty of more than 300% in the processing of the portion of the code to be protected when applied in sequential applications running in single processors. However, this overhead may be less significant when using parallel applications in multicore processors.
- **Conditional Double Modular Redundancy (CDMR)**: In an effort to lower the overhead caused by the TMR technique while maintaining good error detection and correction, we propose a variant of the TMR technique. The CDMR technique executes the protected code twice. If the results of the two executions differ, then the protected code is executed once more. The result of this third execution is then considered a reliable output. This technique hopes to achieve efficiency close to TMR without having to execute three times the same code unnecessarily.

Both TMR and CDMR implementations present time redundancy; that is, the fault tolerance technique itself is not parallel (indeed Sect. 5.4 will present a study on parallel TMR). In the parallel versions of the benchmarks, the parallelism itself is applied to the application algorithm, not to the fault tolerance technique. The fault tolerance techniques presented were applied to the same previously evaluated benchmarks. They are tested using OVPsim-FIM to inject the faults and analyze the errors for each execution. In the following graphs, the percentage of unaces for each application version is presented for each fault tolerance technique and for the unprotected version as a means of comparison of the technique's efficiency. Remember that the unaces value represents faults that did not turn into errors, so high unaces numbers mean excellent protection.

Figure 5.2 presents the graphical results for the vector sum application under fault injection and with the TMR and CDMR techniques applied to it. It is easy to notice that the technique is more efficient for sequential applications. We also see that the fault tolerance techniques show limited improvement in fault mitigation, increasing up to 34.4% the unaces.

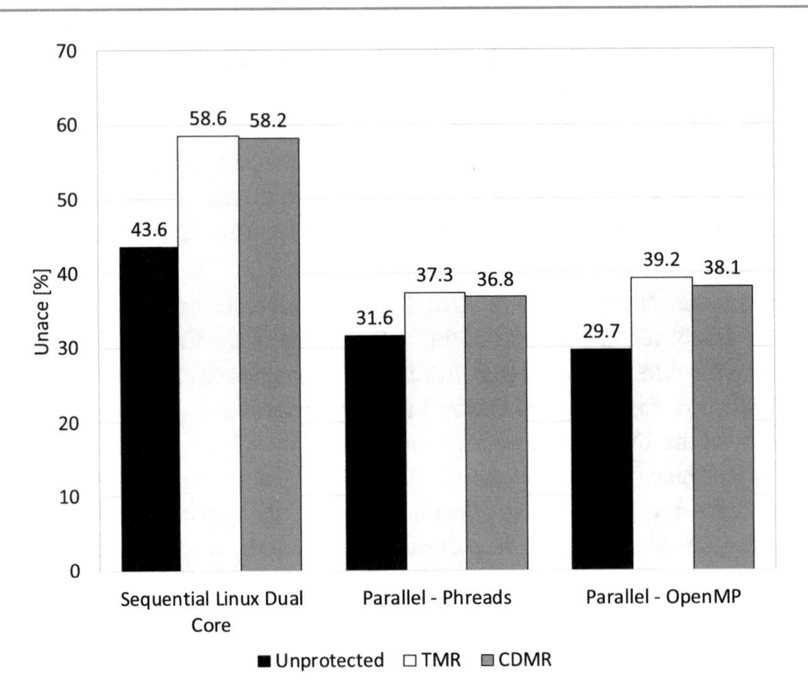

Fig. 5.2 Percentage of unaces on Vector Sum application running on top of Linux OS. *Source* Author

Figure 5.3 presents the results for the matrix multiplication application. In this case, CDMR was less efficient than TMR for the Pthreads execution. Once again, the fault tolerance techniques were not very effective, presenting a maximum increase of only 29.7% on the unaces occurrence. Figure 5.4 presents the results for the bit count application. Again, CDMR was less effective than TMR for the Pthreads version of the application but presented almost the same results for the OpenMP version. In that case, the fault tolerance techniques were even less effective. Figure 5.5 presents the data from the bare metal applications running with and without fault tolerance. It shows that fault tolerance techniques were more effective when executing on sequential bare metal than on Linux (both sequential and parallel). In the best-case scenario, the usage of the techniques has increased by 80.7% the unaces occurrence.

Notice that both TMR and CDMR techniques are not able to mask all the errors. That is expected when executing an operating system apart from the application, which may present its own errors. Our fault tolerance techniques do not protect Linux OS itself, only the applications. In addition, the studied fault tolerance techniques only protect the code from SDCs. That explains the coverage not being 100% on bare metal. An SDC is considered any mismatch in the final memory signature and not only in the application output data memory. Therefore, any SDC error in the memory that does not necessarily affect the application output is considered an error too.

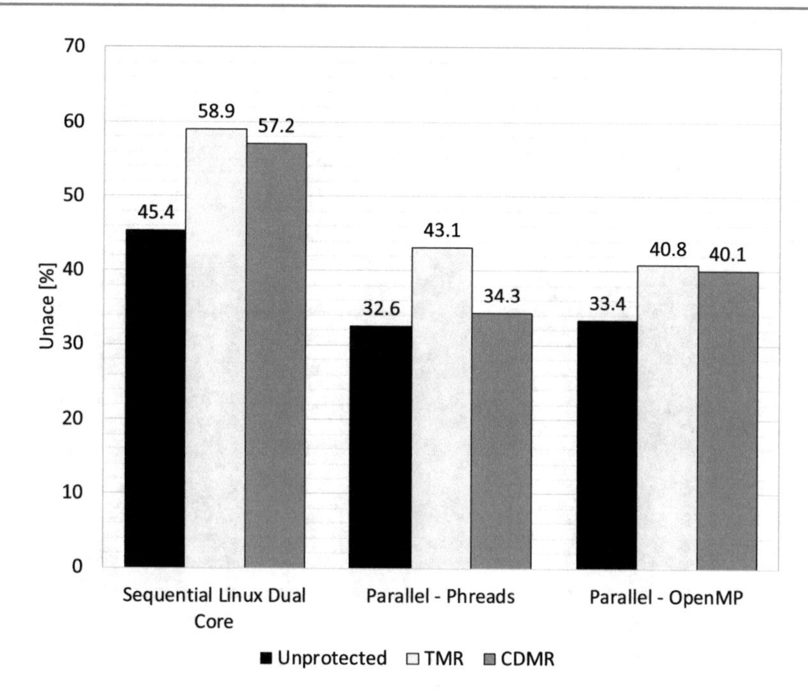

Fig. 5.3 Percentage of unaces on Matrix Multiplication application running on top of Linux OS. *Source* Author

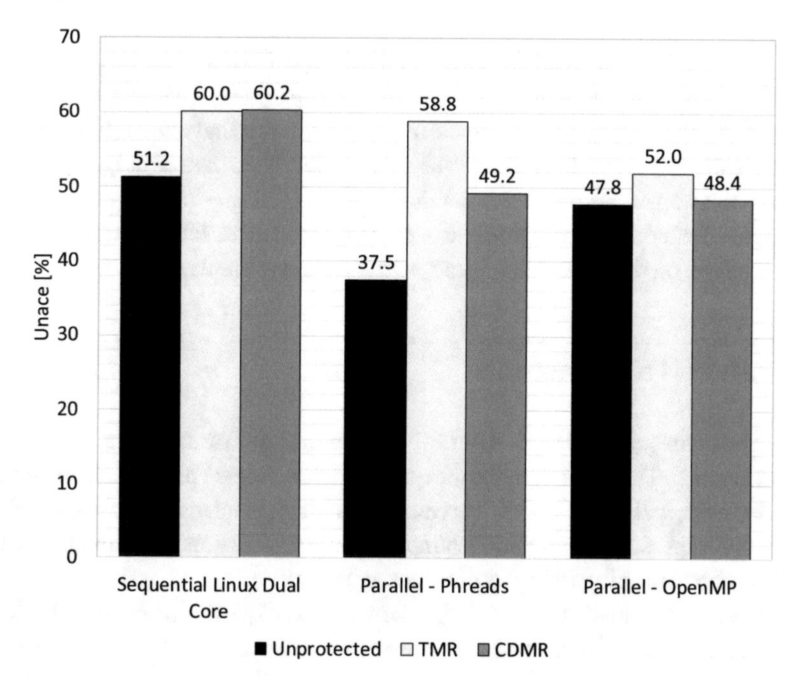

Fig. 5.4 Percentage of unaces on Bit Count application running on top of Linux OS. *Source* Author

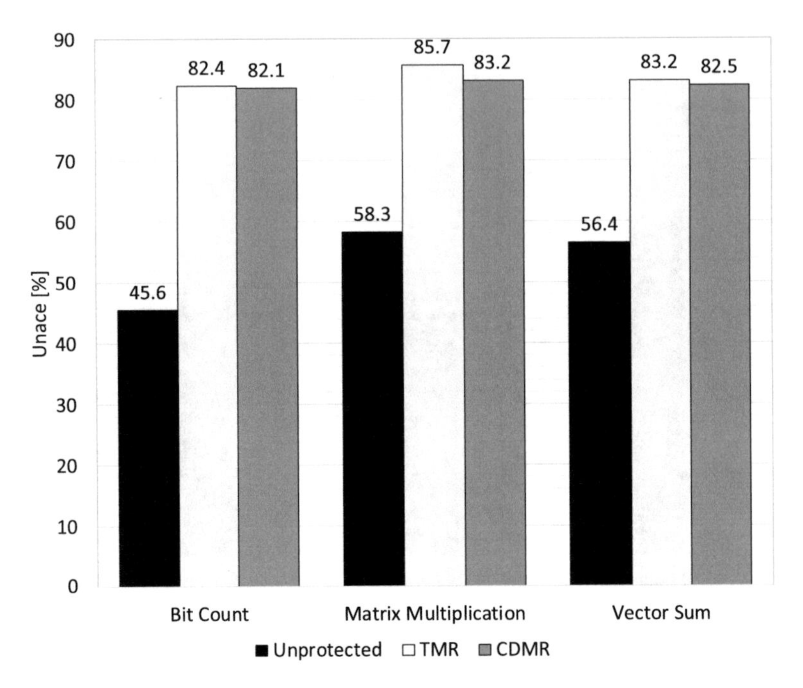

Fig. 5.5 Percentage of unaces on all algorithms running bare metal. *Source* Author

Table 5.3 presents the performance overhead by using TMR and CDMR. Each overhead was calculated with respect to the same algorithm's unprotected code execution time (e.g., an overhead of three means an execution time equals the original one multiplied by three). CDMR presents a lower overhead than TMR, as expected. Note that the usage of an operating system does not impact the overhead, nor does the use of single- or dual-core. Just like the speedups from Table 5.2, the performance overhead presented in Table IV was gathered from the real execution on the Zynq-7000 board, and not from a simulation.

5.4 Parallel Fault Tolerance

The results from the previous section put in evidence the need for fault tolerance techniques on parallel systems. This section raises the question of the impact of parallelism on fault tolerance techniques, and how it varies according to parallelism intensity. It evaluates different fault tolerance approaches with various amounts of threads, stimulating the Linux threads scheduler to obtain data regarding its error susceptibility.

To evaluate the parallelization of fault tolerance techniques, four parallel TMR implementations of a matrix multiplication kernel are proposed: one based on sequential execution and three on parallel execution. The matrix multiplication divides the workload between the

Table 5.3 Performance overhead of fault tolerance techniques

Application	Version	TMR	CDMR
Bit count	Bare metal	3.05	2.01
	Linux sequential	2.87	1.87
	Linux parallel	2.87	1.93
Matrix multiplication	Bare metal	3.00	2.04
	Linux sequential	2.94	1.97
	Linux parallel	3.00	1.99
Vector sum	Bare metal	3.11	2.70
	Linux sequential	2.95	1.98
	Linux parallel	2.98	2.01

two processing cores creating threads during the execution time (using Pthreads). Resource management and thread scheduling are in the hands of the Linux OS scheduler. The total number of threads (i.e., including the matrix multiplication and TMR technique) and available processor cores are expected to influence the application fault tolerance. The approaches are presented below, in order of parallelism intensity:

- **Full Sequential TMR (FS-TMR)**: This implementation is entirely sequential, and as a consequence, the matrix multiplication executes three times voting the result at the end. Notice that the Pthreads API is not used in this implementation.
- **Parallel Execution, Sequential TMR (PE-STMR)**: In the PE-STMR implementation, the algorithm to be protected is executed in parallel, but the TMR technique is coded sequentially. That means the TMR sequentially executes three times a parallel matrix multiplication algorithm, and then the results are voted.
- **Sequential Execution, Parallel TMR (SE-PTMR)**: For this implementation, the TMR technique is parallelized, but the matrix multiplication algorithm is sequential. Therefore, the processor will launch three threads in parallel (TMR redundancies), and a sequential code will be executed on each of these threads.
- **Full Parallel TMR (FP-TMR)**: Finally, both the matrix multiplication application and the TMR technique are implemented as parallel code. Three threads are launched in parallel (TMR redundancies), and each of them executes parallel code.

The PE-STMR adopts a sequential TMR approach with a parallel application. Thus, the primary application flow creates the two child threads, waiting for its completion to launch the next two threads. In contrast, the SE-PTMR implements a parallel TMR into a sequential application (i.e., the matrix multiplication is calculated in a single thread). The FP-TMR deploys a total of nine threads (one per redundancy execution plus two for each one of them to parallel the matrix multiplication). Finally, the sequential version (FS-TMR) employs no parallelism, creating no new threads.

Like in the last section, the testing experiments performed on the techniques were the fault injection simulation with the OVPSim presented in Sect. 4.5, with the same types of errors defined at Table 4.1. However, a new classification is added: the "Masked" data presented in the following figures represent the percentage of errors that were present in the system but were masked by the TMR techniques (they would become SDCs if no TMR was implemented).

Figure 5.6 presents the fault masking performances for the executions under fault injection. It is noticeable that about 35% of the FS-TMR occurrences are unaces, that is, no error at all. Also, it was capable of masking about 50% of the SDCs. Results show that PE-STMR achieved much more fault tolerance. Only about a fourth of the SDCs were not corrected by it. SE-PTMR result data is similar to the results from PE-STMR. The fault tolerance is lower in this implementation, however. FP-TMR results are also very similar to PE-STMR. Comparing the FP-TMR and the SE-PTMR with the PE-STMR approach, the last one achieves better fault tolerance. It indicates that parallelizing the TMR technique can have a negative impact on dual-core processors, augmenting the number of errors. The fact that FP-TMR is the one with the less occurrence of unaces and corrected errors shows that the Pthreads API is terrible for fault tolerance. Nevertheless, the parallel TMR benchmarks showed better performance, indicating that the independence between threads confines the error in a single Pthread context (which is only partially shared).

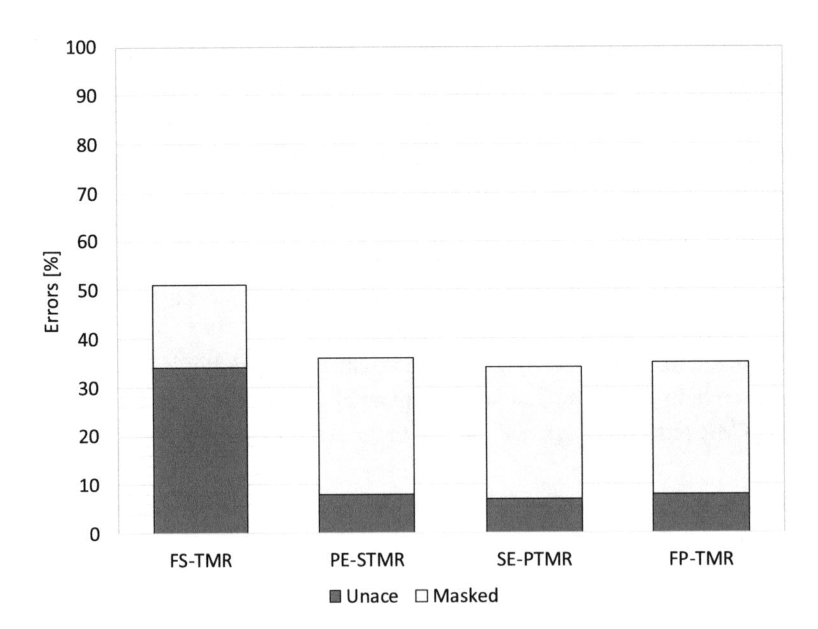

Fig. 5.6 Comparison between all the TMR fault masking performances. *Source* Author

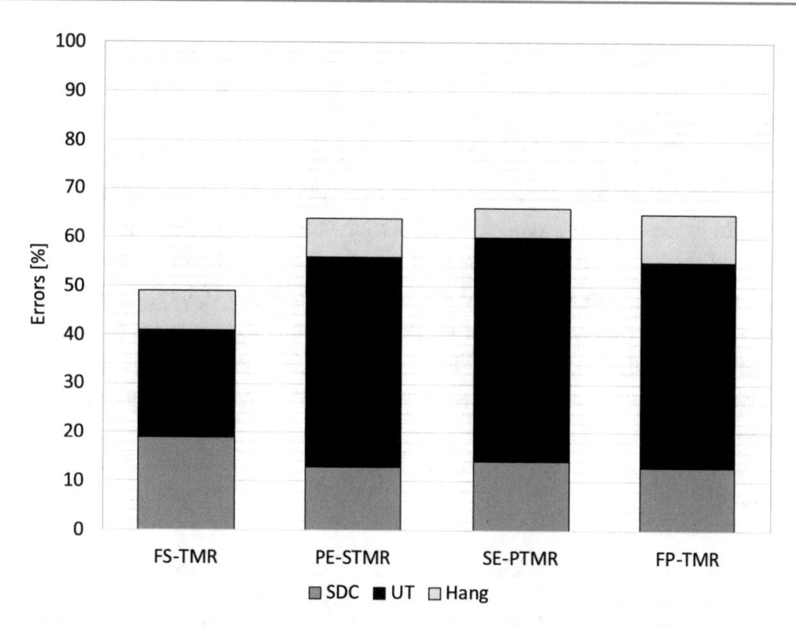

Fig. 5.7 Comparison between all the implementation error occurrences. *Source* Author

Figure 5.7 presents a comparison between all the benchmark error occurrences. An interesting fact is that the full sequential code (FS-TMR) shows about half the percentage of unexpected terminations (UT) when compared to parallel applications. Unexpected terminations are a type of error that occurred during the testing of parallel algorithms: it means that the algorithm was terminated by the operating system before completing its expected execution. Also, no matter the number of threads, all parallel applications are equally susceptible to this type of error. PE-STMR is the technique less vulnerable to SDCs, while SE-PTMR is the most susceptible among the parallel ones. This is another indication that parallelizing TMR has a severe impact on fault tolerance. On the other hand, parallelizing the application reduces the susceptibility to SDCs.

5.5 Conclusion

It is clear that fault tolerance techniques shall be carefully developed and chosen with the specificity of the system to be protected in mind. As the previous sections showed, even different implementation strategies of the same fundamental fault tolerance technique can have very different performances regarding fault masking capability and execution time cost. We have chosen to evaluate two more traditional techniques (CDMR and TMR) extensively instead of testing out a plethora of different ones because that is how the development of a dependable system goes. Usually, the fault tolerance method used is one that has been standing the test of time, like TMR. When developing safety-critical systems, there is little room for taking chances. Thus, the work concentrates on giving the chosen fault tolerance method the best performance possible.

The next chapter will present approximate computing methods and how they can be used to provide intrinsic fault tolerance to a system and approximate our traditionally used fault tolerance techniques. Approximate fault tolerance is less costly than its not approximated counterparts. They also are, more often than not, able to achieve almost the same fault detection and masking numbers. Over the final part of the book (Chap. 7), the fault tolerance techniques studied in this chapter will be further evaluated, being put to stress test under real case scenario emulation. They will be run under fault injection, and their fault tolerance will be assessed so that we can compare them with their approximate counterparts.

References

1. A. Avizienis, J. Laprie, B. Randell, C. Landwehr, Basic concepts and taxonomy of dependable and secure computing. IEEE Trans. Dependable Secur. Comput. **1**(1), 11–33 (2004)
2. J.R. Azambuja, M. Altieri, J. Becker, F.L. Kastensmidt, Heta: Hybrid error-detection technique using assertions. IEEE Trans. Nucl. Sci. **60**(4), 2805–2812 (2013)
3. L. Chen, A. Avizienis, N-version programminc: a fault-tolerance approach to reliability of software operation, in *Twenty-Fifth International Symposium on Fault-Tolerant Computing, 1995, ' Highlights from Twenty-Five Years'* (1995), p. 113
4. E. Chielle, R.S. Barth, Â.C. Lapolli, F.L. Kastensmidt, Configurable tool to protect processors against see by software-based detection techniques, in *2012 13th Latin American Test Workshop (LATW)* (2012), pp. 1–6
5. E. Chielle, G.S. Rodrigues, F.L. Kastensmidt, S. Cuenca-Asensi, L.A. Tambara, P. Rech, H. Quinn, S-seta: selective software-only error-detection technique using assertions. IEEE Trans. Nucl. Sci. **62**(6), 3088–3095 (2015)
6. E.P. de Freitas, M.A. Wehrmeister, E.T. Silva, F.C. Carvalho, C.E. Pereira, F.R. Wagner, *DERAF: A High-Level Aspects Framework for Distributed Embedded Real-Time Systems Design* (Springer, Berlin, 2007), pp. 55–74
7. A.B. de Oliveira, G.S. Rodrigues, F.L. Kastensmidt, N. Added, E.L.A. Macchione, V.A.P. Aguiar, N.H. Medina, M.A.G. Silveira, Lockstep dual-core ARM A9: implementation and resilience analysis under heavy ion-induced soft errors. IEEE Trans. Nucl. Sci. **65**(8), 1783–1790 (2018)
8. G. Dong, N. Xie, T. Zhang, On the use of soft-decision error-correction codes in NAND flash memory. IEEE Trans. Circuits Syst. I: Regul. Pap. **58**(2), 429–439 (2011)
9. M. Fayyaz, T. Vladimirova, Fault-tolerant distributed approach to satellite on-board computer design, in *2014 IEEE Aerospace Conference* (2014), pp. 1–12
10. O. Goloubeva, M. Rebaudengo, M. Sonza Reorda, M. Violante, Soft-error detection using control flow assertions, in *Proceedings 18th IEEE Symposium on Defect and Fault Tolerance in VLSI Systems* (2003), pp. 581–588
11. V. Hilderman, T. Baghi, Avionics certification: a complete guide to DO-178 (software), DO-254 (hardware). Avionics Communications (2007)
12. B. Huang, R. Sass, N. Debardeleben, S. Blanchard, Harnessing unreliable cores in heterogeneous architecture: the PyDac programming model and runtime, in *2014 44th Annual IEEE/IFIP International Conference on Dependable Systems and Networks* (2014), pp. 744–749

13. K.-H.Huang, J.A. Abraham, Algorithm-based fault tolerance for matrix operations. IEEE Trans. Comput. *C-33*(6), 518–528 (1984)
14. P. Koopman, T. Chakravarty, Cyclic redundancy code (CRR) polynomial selection for embedded networks, in *International Conference on Dependable Systems and Networks* (2004), pp. 145–154
15. J.S. Monson, M. Wirthlin, B. Hutchings, Fault injection results of linux operating on an FPGA embedded platform, in *2010 International Conference on Reconfigurable Computing and FPGAs* (2010), pp. 37–42
16. H. Mushtaq, Z. Al-Ars, K. Bertels, Efficient software-based fault tolerance approach on multicore platforms, in *2013 Design, Automation Test in Europe Conference Exhibition (DATE)* (2013), pp. 921–926
17. N. Oh, S. Mitra, E.J. McCluskey, ED4I: error detection by diverse data and duplicated instructions. IEEE Trans. Comput. **51**(2), 180–199 (2002)
18. H. Quinn, Z. Baker, T. Fairbanks, J.L. Tripp, Software resilience and the effectiveness of software mitigation in microcontrollers. IEEE Trans. Nucl. Sci. **62**(6), 2532–2538 (2015)
19. H. Quinn, Z. Baker, T. Fairbanks, J.L. Tripp, G. Duran, Robust duplication with comparison methods in microcontrollers. IEEE Trans. Nucl. Sci. **64**(1), 338–345 (2017)
20. A.J. Sanchez-Clemente, L. Entrena, M. Garcia-Valderas, Partial TMR in FPGAs using approximate logic circuits. IEEE Trans. Nucl. Sci. **63**(4), 2233–2240 (2016)
21. T. Santini, L. Carro, F. Rech Wagner, P. Rech, Reliability analysis of operating systems and software stack for embedded systems. IEEE Trans. Nucl. Sci. **63**(4), 2225–2232 (2016)
22. L. Sterpone, M. Violante, An analysis of SEU effects in embedded operating systems for real-time applications, in *ISIE 2007. IEEE International Symposium on Industrial Electronics, 2007* (IEEE, 2007), pp. 3345–3349
23. L.A. Tambara, P. Rech, E. Chielle, J. Tonfat, F.L. Kastensmidt, Analyzing the impact of radiation-induced failures in programmable SoCs. IEEE Trans. Nucl. Sci. **63**(4), 2217–2224 (2016)

6 Approximate Computing and Fault Tolerance

6.1 Summary

Chapter 2 presented a multitude of approximation methods. This chapter will put into practice the ideas presented before and discuss their real impact on the system's accuracy and costs. The chapter is divided into two sections. Section 6.2 proposes three approximate computing techniques intended for general use. This section presents those techniques and a brief evaluation of their applicability and cost. All those approximation strategies can be applied both to programmable hardware and embedded software. Section 6.3 discusses the general use of approximation on fault tolerance and some of its motivations. Section 6.4 uses two of the proposed approximate computing techniques and another one introduced in Chap. 2 to develop approximate versions of a traditional fault tolerance technique: the triple modular redundancy. Section 6.5 proposes an approximate error detection technique developed for multicore real-time systems.

6.2 Approximation Methods

This section will cover in practice three approximation methods: data precision reduction, functional approximation, and loop-perforation. Section 6.2.1 starts by evaluating data precision reduction implemented in hardware and how it can be used to reduce the used area. Section 6.2.2 presents a combination of functional approximation and loop-perforation applied. Finally, Sect. 6.2.3 proposed a numerical specific approximation that motivates the universal use of approximate computing.

G. S. Rodrigues et al., *Approximate Computing and its Impact on Accuracy, Reliability and Fault-Tolerance*, Synthesis Lectures on Engineering, Science, and Technology,
https://doi.org/10.1007/978-3-031-15717-2_6

6.2.1 Data Precision Reduction

Data precision reduction approximation can be easily implemented on programmable hardware. This is done by creating new data types with reduced bit-size. Representing values with a limited number of bits saves hardware resources in the detriment of data precision. As will be detailed further, high-level synthesis (HLS) is used to generate hardware from software code; thus, the proposed approximation method is assured to be applicable both to software and hardware projects, although the impact on each would be different.

Variables on software are usually defined by standard types. Those types define how the variables' read, write, and arithmetic and boolean operations are executed on the hardware, as well as their bit-size. All those characteristics have a significant impact on the system performance, energy consumption, and also fault tolerance [3]. The same is valid for variables on projects that make use of hardware description languages (HDL), specifically when those are generated by HLS.

The benchmark designs analyzed in this section are coded in C language. The hardware implementation is then generated by Vivado HLS. Vivado HLS is a tool provided by Vivado Design Suite (Xilinx). It produces hardware description using a C or C++ language code as input. Vivado HLS also counts on a series of hardware optimizations and generation tools, providing considerable control over the final product. The *ap_fixed.h* library provided by Vivado HLS allows arbitrary data type creation, for fixed-point variables, and is used in this work to create approximate data types. In the regular floating-point representation (IEEE-754 standard), 28% of the 32 bits are designated to represent the exponent while the other 72% represents the significand. The approximate fixed-point data types evaluated in this section keep that same share for the representation of the numbers above the decimal point and the value below the decimal point.

Figure 6.1 presents the used area—concerning DSPs—for eight floating-point data type sizes applied to both operands of a simple multiplication between two variables implemented in the FPGA of a Zynq-7000 APSoC by Xilinx (further implementation details are given at Chap. 7). The figure shows that the multiplication between two traditional IEEE-754 standard *float* variables costs one less DSP than using 32-bit variables implemented by the *ap_fixed.h* library. Nevertheless, variables generated by *ap_fixed.h* with less than 32 bits consume less DSPs. This alone is an indication of its capability of saving resources that can then be used to improve performance. In this particular case, the saved DSPs could then be used to improve performance, executing other operations in parallel. Figure 6.1 shows that it is possible to implement two 28-bit multiplications in parallel using the same number of DSPs as one single 32-bit operation. Data precision reduction implemented on embedded software will also be studied in this work. As Sect. 6.4.2 will discuss, the effects on software are different. While it affects the usage of programmable hardware, on embedded software, it will have a direct influence on the memory footprint of the application.

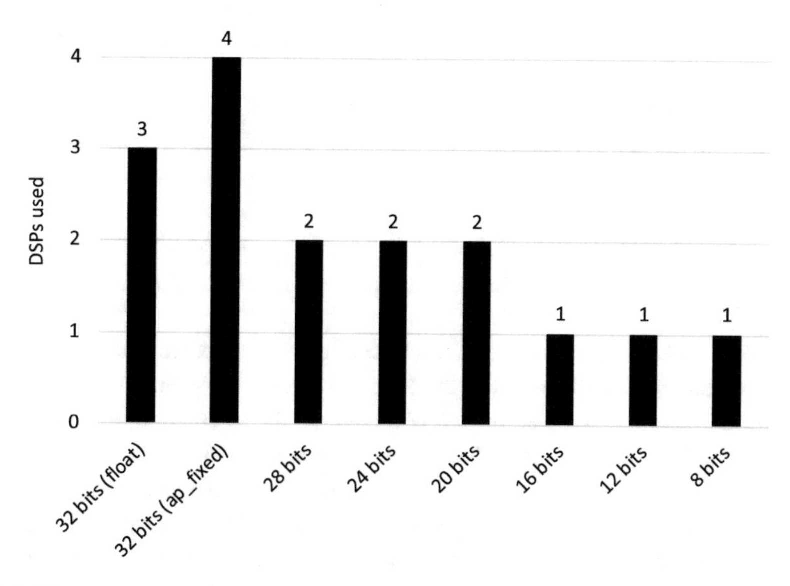

Fig. 6.1 DSP usage for multiplication using different data type configurations to represent floating-point values. *Source* Author

6.2.2 Successive Approximation

This proposal consists of a type of functional approximation combined with loop-perforation. This method is an excellent example of how two theoretical ideas for approximation can be combined. It consists of using an inherent characteristic of the algorithm, namely its loop-based execution, in favor of approximation. As will be discussed, this study case also shows how some algorithms are inherently approximate, and that some solutions (such as numerical calculations) can only be solved approximately.

Successive approximation algorithms are numerical calculations for which an exact, straightforward solution is simply not computationally achievable. Such is the case with the calculation of the integral of a function. Those algorithms are iteration-based and get closer to an acceptable result on every iteration. An example of this kind of numerical algorithm is the trapezoidal rule, which is used to calculate the area under a curve approximately (i.e., the integral of $f(x)$) as the sum of trapezoid areas, as defined in (6.1).

$$\int_{a}^{b} f(x)dx \approx (b-a)\left(\frac{f(a)+f(b)}{2}\right) \tag{6.1}$$

One can take (6.1) and make a more accurate approximation by breaking up the interval between the points a and b into a number n of smaller intervals. Then, the algorithm consists of computing the approximation for each of those intervals and adding the results afterwards, achieving a better result than the one provided by (6.1) alone. The bigger the size of n, the better the result will approximate the real integral solution. This is called an iterated rule. The iterated calculation for the trapezoidal rule is presented in (6.2). In this case, the intervals present the form $[kh, (k+1)h]$, where $h = (b-a)/n$ and k ranges from 0 to $n-1$.

$$\int_a^b f(x)dx \approx \frac{(b-a)}{n}\left(\frac{f(a)}{2} + \sum_{k=1}^{n-1}\left(f\left(a + k\frac{b-a}{n}\right)\right) + \frac{f(b)}{2}\right) \tag{6.2}$$

Because the value is approximated in each iteration, it is expected that if an error occurs, causing an iteration result that is out of the expected calculation path, it will be corrected in the following iterations, as the algorithm will then get the calculation back on track. This statement is valid when analyzing soft errors, not permanent errors. In the case of permanent errors that compromise the hardware and the right execution of the algorithm, the executions of all future iterations would be compromised, therefore compromising the convergence of the algorithm. Moreover, if a SEU occurs in one of the last iterations (or the very last one) it may be too late for the algorithm to converge back to an acceptable result. Therefore, it is expected that the higher the value of n, less susceptible to errors the algorithm will be. Conversely, having a higher number of n would also increase the execution time of the computation. Past works show that applications executing in a radioactive environment with a higher execution time are more prone to have errors, as they would be exposed to more radiation [11] and TID [10].

Increasing the number of n would, therefore, have a positive impact on the algorithm's protection against faults and exactitude (the higher the iteration number, the higher the accuracy). However, it will negatively impact the execution time performance. This negative impact on performance could then degrade the system reliability, even to the point of repealing the positive impact from the successive approximation. Intuitively, the higher the execution time, the higher the exposure time, so the probability of observing faults leading to failures increases with n. Those factors shall be intensely studied by a developer that intends to use numerical methods on safety-critical systems. As stated before, some solutions are only achievable by numerical analysis, so this study is imperative for any complex critical-system development.

Because of their nature, successive approximation algorithms and numerical methods, in general, have their own sources of errors (manifested as inaccuracy):

- **Simplification Errors**: because every numerical method is indeed an approximate model of reality, they can only relate to the mathematical reality to a certain extent.
- **Truncation Errors**: given that the accuracy of floating-point values is limited, exactitude may be lost.

- **Accumulation Errors**: in some numerical algorithms errors may propagate, so the final result will be less exact than middle-term results. The types of errors above may also contribute to the accumulation of errors on each iteration.

The first two types of errors are also present in ordinary computation methods, but the last one is natural for iterative algorithms, like numerical methods. It is important to keep in mind, however, that this kind of error is different from the one caused by SEUs. The former is a characteristic of the programming paradigm, while the latter is caused by a harmful environment. Programming paradigms are known to affect the fault tolerance of a system even when using numerical methods are not being used.

Another critical factor to take into account when using this type of computational method is convergence. It is possible for some numerical methods to converge faster to the result with the required exactitude, thus requiring a lower number of n iterations. It is also possible for some methods to require a higher number of n. The real problem, however, is when the method diverges. Some methods can even converge for a specific interval and diverge for another. The convergence problem is out of the scope of this book, but it is important to point out that as an issue. There are, fortunately, numerical methods that are guaranteed to always converge to a correct, acceptable result.

6.2.3 Taylor Series Approximation

Given the extensive amount of possible approximate computing methods presented by the literature, finding the most suitable strategy for a given application is a significant challenge. Most approximation methods presented in the literature are specially developed for a single application, being unscalable, and many times even inapplicable for a different purpose or algorithm. Theoretically speaking, one can say that infinity of applications has still not been approached by approximate computing studies, and never will. For some applications, developing a unique approximate design might be extreme intellectual work.

One of the proposals of this book is to use the Taylor series to numerically approximate functions. Although many numerical techniques for mathematical approximation are available, the Taylor series was selected due to its high applicability and simplicity [12]. Another reason to choose it was the fact that its terms can be previously calculated, implying a performance gain when implemented in software. In particular, the Taylor series was chosen instead of the Maclaurin series due to its higher usability and generality [5, 9]. Maclaurin series is a specific case of the Taylor series and has smaller applicability due to its constraints.

Because mathematical functions can represent any algorithm, this approximation approach can be used to generate an approximate version of almost any given software (given some mathematical limitations imposed by the technique, to be further defined). The same strategy can also be used to provide approximate hardware. Both software and hardware implementations are better discussed in Chap. 7. The main contribution of this part of the work is to provide a theoretical basis for the generation of approximate numerical versions of any software or hardware design. This proposal relies on approximation theory and mathematical analysis to provide mathematically valid approximations.

Taylor series are used in mathematics to represent a function as a sum of previously calculated terms. These terms are generated from the values of the function's derivatives at a given point. The more the terms are used, the better is the representation. This way, functions are approximated using a finite number of terms in a Taylor series. An infinite number of terms would adequately represent the original function. However, calculating infinite terms is computationally impossible. The compact sigma notation for Taylor series is presented in (6.3), where n stands for the current term (from 0 to N), and a stands for the center point (where the derivatives are calculated), being $f^{(n)}(a)$ the nth derivative of the function f at the point a. When $a = 0$, the Taylor series is called a Maclaurin series.

$$\sum_{n=0}^{N} \frac{f^{(n)}(a)}{n!}(x-a)^n \tag{6.3}$$

Being an approximation, a Taylor series with finite terms presents a quality degradation when compared with the original function. This degradation is variable depending on the center point used and the number of terms. For some functions, the Taylor series may converge in a given range and diverge when out of its bounds. However, it is possible to estimate this quality loss quantitatively using Taylor's theorem. Functions that contain one or more singularities cannot be represented as Taylor series either. The convergence of a given function using Taylor series approximation needs to be evaluated before its usage. The designer is also accountable for employing a sufficient number of Taylor terms so that the quality loss does not interfere with the generation of a *good enough* result.

Some of the approximation problems presented by the Taylor series can be dealt with by the designer during implementation time. For instance, if the range of the input values of a function is known, the designer can evaluate if an approximation by the Taylor series is feasible. Also, even when such an approximation shows up as divergent in a critical range, the designer can achieve a good approximation by merely changing the Taylor series center point a (see Eq. 6.3).

As previously discussed in Sect. 6.2.1, approximate computing can be used to reduce the size of programmable hardware implementations and improve the execution time of software. Both hardware size and execution time are highly related to reliability, therefore using the Taylor series to approximate any given function is expected to improve their fault tolerance as well. Thus, varying the number of Taylor series terms will impact not only resource usage but also tolerance. For those reasons, this study is fundamental for safety-critical system designers willing to implement numerical algorithms.

An approximation for the exponential function is implemented as a benchmark to test the proposed Taylor series approximation. The algorithm is coded in C and implemented both in software and hardware. Hardware implementations are generated by Vivado HLS using the very same code and implemented in the FPGA of a Zynq-7000 APSoC. The software

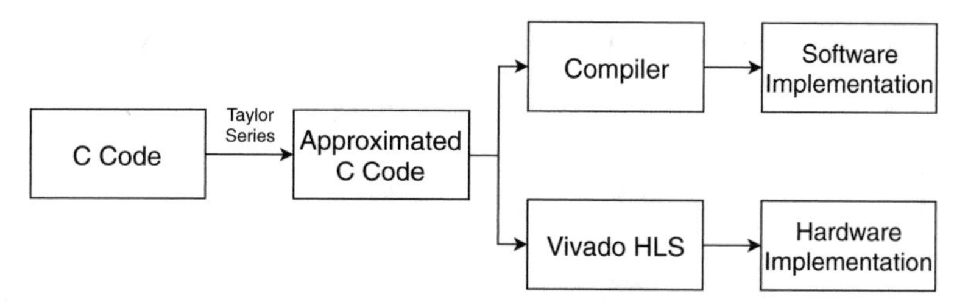

Fig. 6.2 Taylor series approximations implementation flow. *Source* Author

versions are executed in the embedded ARM A9 processor embedded on the same board. Figure 6.2 presents the implementation diagram of the approximations. The mathematical equation for this Taylor series approximation is presented in (6.4).

$$e^x = \sum_{n=0}^{N} \frac{x^n}{n!} = 1 + x + \frac{x^2}{2!} + \frac{x^3}{3!} + \cdots + \frac{x^N}{N!} \tag{6.4}$$

Hardware Implementation Two parameters are considered during the hardware implementation: data precision type and the usage of a pipeline. The data precision types evaluated are *double* and *float* (i.e., double-precision and single-precision floating-point formats, with 64 and 32 bits, respectively). Pipeline is used to accelerate the Taylor series computation loop and is implemented into the algorithm by merely adding a *pragma* (option from Vivado HLS) in the C code. No particular argument is passed to the HLS pipeline pragma, so by default, it will try to pipeline the loop as much as possible. Each possible combination of those two parameters is tested, resulting in four implementations (i.e., *double* and *float* variables, with and without pipeline).

Table 6.1 presents the resource usage for the four implemented Taylor series approximations. It shows the data from the four variants of the approximation for a variety of numbers of Taylor series terms. The double-precision variant with pipeline presents data from 3 to 13 terms because it is the maximum number of term implementations that fits in the Zynq-7000 FPGA. For the same reason, the float precision variant with the pipeline is presented from 3 to 34 terms. Resource usage concerning area is divided into four categories: DSPs, FFs, LUTs, and Essential Bits. The last two columns present the data for latency (in clock cycles) and accuracy (in percentage). The accuracy is calculated by comparing the output value with the best value obtainable computationally without Taylor series approximation (the exponential function from the *math.h* C library for the given data precision). Vivado HLS implementation reports provided the estimation of the hardware latency and area resource usage presented in the table.

It becomes clear by the analysis of Table 6.1 that the usage of pipeline profoundly affects area resource occupation, while the latency slightly increases with the number of terms. On the other hand, the absence of pipeline implies an almost constant hardware area but provokes an enormous latency increase with a higher number of terms. The essential bits

Table 6.1 Performance and resource usage analysis from HLS hardware implementation of Taylor series approximation

Precision	Pipe.	Terms	Area				Lat. [c.c.]	Accuracy [%]
			DSPs	FFs	LUTs	Esst. Bits		
Double	No	5	14	4761	6250	804046	215	89.28417678
		10	14	4765	6254	824732	430	99.97332604
		25	14	4769	6258	810620	1075	100
		50	14	4773	6263	825134	2150	100
		100	14	4777	6267	806337	4300	100
	Yes	3	28	1797	3518	449824	16	54.70235457
		4	42	5987	8967	1126115	41	76.02448002
		5	67	7264	11400	1409144	47	89.28417678
		8	109	19777	27747	3502632	65	99.58761712
		11	162	29396	41078	5183543	83	99.99410196
		13	190	37738	51976	6618716	95	99.99977405
Float	No	5	5	1648	2361	251843	130	89.28417875
		10	5	1652	2365	255535	260	99.97332926
		25	5	1656	2369	250510	650	99.99998919
		50	5	1660	2374	258650	1300	99.99998919
		100	5	1664	2378	256350	2600	99.99998919
	Yes	3	10	837	1454	168274	12	54.70235538
		4	15	2054	3191	342306	24	76.02448475
		5	23	2646	4255	445723	29	89.28417875
		8	38	6276	9498	991407	44	99.58761582
		16	81	15338	22753	2377350	84	99.99998919
		34	177	35882	52737	5494006	174	99.99998919

are configuration bits that are really used by the design, on which a bit-flip will possibly cause errors. This data is important for safety-critical system design, where it shall be as low as possible. The table also shows that double-precision (*double*) achieves accuracy per number of terms almost at the same rate as single-precision (*float*). Nevertheless, only double-precision was capable of achieving full accuracy. Another interesting fact is that not many terms are needed to provide good accuracy. In fact, 8 terms seem to be enough to provide an accuracy of 99% for any implementation. From that point further, the area and latency costs increase, but the accuracy remains almost the same.

Software Implementation: The same code used at Vivado from the last section was also implemented on Vivado SDK and executed on the ARM processor. The software is bare metal implemented. However, in this case, only two versions of the algorithm are presented: one for double-precision and another for single-precision. That is because there is no implementation strategy on embedded software equivalent to the Vivado HLS pipeline.

Table 6.2 presents the data from the embedded software execution performance. The two columns present the data for execution latency (in clock cycles) and accuracy (in percentage).

Table 6.2 Performance analysis from embedded ARM software implementation of Taylor series Approximation

Terms	Double		Float	
	Exec. Lat.[c.c.]	Accuracy[%]	Exec. Lat.[c.c.]	Accuracy[%]
2	230	28.9872284	238	28.98722979
3	202	54.70235457	164	54.70235531
4	252	76.02448002	204	76.02448465
6	348	95.88087592	284	95.88087586
8	444	99.58761712	364	99.58761569
10	540	99.88980474	444	99.97332912
13	684	99.99977405	576	99.99978126
15	780	99.99999351	644	99.99999681
16	828	99.99999986	696	99.99998933
20	1020	100	844	99.99998933

The accuracy is calculated by comparing the output value with the best value obtainable for the given data precision (using the function from the *math.h* library). The execution latency of the embedded software is measured by executing the applications on the ARM Cortex-A9 processor embedded Zynq board making use of the *xtime_l.h* C library provided by Xilinx.

As expected, both the execution latency and accuracy increased with the number of Taylor series terms. Table 6.2 shows that the accuracy increases exponentially with the increase of the number of terms. The latency also increases with the number of Taylor series terms, but not as much. Surprisingly, single-precision appears as a better choice when using 10 to 15 terms, providing both better accuracy and execution latency.

6.3 Approximate Fault Tolerance: Discussion and Motivations

Approximation itself implies the idea of inherent error tolerance. On approximate systems, a specified error tolerance has to be considered, but that is not the same error definition used when discussing radiation effects and safety-critical systems. Approximation errors are caused by the system itself and manifested as quality or accuracy degradation. Also, when dealing with approximation, the decision of whether an error caused a failure or not is a matter of definition related to what would be considered a "correct" application output, which is often hard to be defined. Taking, for instance, the example of image outputs, the correctness of the output is tied to an image quality definition, which is different from one human being to another because of biological reasons. This accuracy relaxation from the approximate system can, however, be used in favor of fault tolerance on safety-critical systems: a system that accepts some accuracy degradation can ignore errors in memory that have a low impact on the data value, for example. Also, the reduction of the complexity, achieved by approximation, can help to reduce the system's susceptibility to faults (e.g., by reducing the critical area of a hardware circuit).

In safety-critical systems, however, the definition of error is related to the occurrence of a fault. In this scope, the approximation can be used in two manners. First, it can be used to improve the application execution time, energy consumption, and even reliability. Secondly, approximate computing can also be used to reduce the costs of fault tolerance techniques. The impact of using approximation on those two levels, however, is different. As already discussed, the approximation of the application directly impacts its accuracy, and therefore reliability. Approximating fault tolerance techniques may, however, be developed in such a way as to avoid affecting the accuracy of the application, or affecting it only up to the acceptable level that is defined by its quality (or accuracy) requirements.

TMR is one of the most traditional fault tolerance techniques presented by the literature. Approximate TMR (ATMR) [7] is based on implementing each redundancy task with a different architecture or algorithm to provide the capability of masking multiple errors. When applied to hardware projects, ATMR has been presented as a way to achieve fault coverage almost as good as traditional TMR but avoiding the huge area overhead that it costs [2]. Designers might accept a lower fault coverage if the area overhead of the project is to drop significantly. Also, a smaller hardware area implies higher fault tolerance due to the reduction of the critical area. Therefore ATMR might be, in some cases, not only less costly but also more reliable than traditional TMR. In traditional TMR, at least two redundancies need to have the same correct value at a given time so that the correct output can be voted. Using approximations on TMR is not trivial, because of the errors caused by the accuracy loss: even in the absence of a fault, two TMR redundancies of different accuracies will present different outputs. In [8], the authors present an ATMR approach that guarantees that the result of at least two redundancy circuits will always be the same (in the absence of a fault). The idea is to use different forms of approximation on each redundancy so that two of them will not be affected by approximation errors at the same time, and the ATMR will be able to mask that error. The authors present their approximation method and prove mathematically that the errors introduced by the approximation will not harm the normal behavior of the ATMR. They also propose a full ATMR (FATMR) approach where all three circuits are approximations (instead of having one non-approximate circuit and two approximations). This ATMR technique can also be used alongside tools that generate the best possible approximate functions with genetic algorithms [1]. The evolutionary algorithm is capable of generating the best combination of approximate functions possible for a given system. However, the ATMR and FATMR methodologies are still limited by their mathematical and theoretical constraints.

Most of the approximation techniques presented in the literature are application-specific. Therefore, it is very hard or impossible to apply the same approximation technique to any possible design or code. Knowing all the possible approximation methods and which type of design is a better fit for each of them is barely impossible work. Also, some approximation methods are applicable to multiple types of applications and hardware designs. Therefore, the designer should test all of them before deciding on the one with better performance. All that would demand design time that most developers cannot afford. This book tries to solve those issues by presenting easy-to-implement approximation methods that can be applied both to programmable hardware and embedded software. Approximate fault tolerance techniques are also proposed by applying those methods to traditional TMR.

6.4 Approximate Triple Modular Redundancy (ATMR)

Given the proposed approximate computing methods, we believe that some of them can be used to improve traditional fault tolerance methods. The most classical fault tolerance method presented in the literature is TMR, as already discussed. Therefore, to evaluate how approximate computing can improve fault tolerance methods, two approximate TMR (ATMR) techniques are proposed, based on two of the approximation techniques presented in this chapter.

Fault tolerance techniques often introduce a high execution time or hardware area overhead. Such is the case of TMR, which costs an overhead of at least 200%. This book proposes an ATMR method to deal with that issue without highly compromising fault tolerance. Different from [1, 2, 8], the ATMR techniques proposed in this work do not work with approximations limited by a mathematical statement. The ATMRs presented in this chapter deal with the concept of *approximation intensity*, where a function can be more (or less) accurate, having a direct impact on the method fault coverage, the final answer accuracy, and the application execution time.

6.4.1 Hardware ATMR Based on Data Precision Approximation

The proposed ATMR benefits from the data precision approximation to generate redundancies that are less accurate than the classical ones, but smaller in area. This ATMR is expected to achieve fault tolerance close to the traditional ones, but with less area overhead. The ATMR is applied to simple codes (two matrix multiplication algorithms). This is intended to evaluate how the studied type of approximation affects data operations and its effects on hardware. Using a sophisticated code could mask that information. The fault tolerance of the proposed technique is assessed with fault injection on the FPGA configuration memory.

Results from Fig. 6.1 prove that the proposed data precision reduction approximation saves resources. This indicates that the proposed approximation can be used to provide an ATMR design with a lower area overhead. If that turns out to be true, it may even be possible to improve general-use designs, achieving better performance and resource usage, as well as fault tolerance (given the lower hardware area).

Listing 6.1 Simplified pseudo-C Code for a an ATMR implementation using 24-bit variables for Vivado HLS.

```
#include <ap_fixed.h>
typedef ap_fixed<32,9> tsize_32;
typedef ap_fixed<24,7> tsize_24;

void main(tsize_32 input_A[2][2], tsize_32 input_B[2][2],
tsize_32 output[2][2])
{
    tsize_24 result1[2][2], result2[2][2], result3[2][2];
    result1 = matrixMult1(matrixA, matrixB);
    result2 = matrixMult2(matrixA, matrixB);
    result3 = matrixMult3(matrixA, matrixB);
    output = bitwiseVoter(result1, result2, result3);
}
```

Listing 6.1 presents a pseudo-C code that summarizes the ATMR implementation. Vivado HLS is used to implement hardware based on them, as explained in Sect. 6.2.1. Some less important parts of the code are left out for simplification purposes. The ATMR is implemented as three operations, in different functions at the C code, so that Vivado HLS is forced to implement specific hardware for each one of them. Otherwise, it could reuse hardware, which is not desired for the TMR implementation. The voter is implemented as a single independent function and consists of boolean operations that perform a bitwise check between three values.

Between the matrix multipliers and the voter, converters may or may not be needed: depending on the sizes of the data in use. That is because the voter cannot vote for values of different bit-sizes. Converters may also be needed inside the matrix multiplier function implementation, in case the input matrices are of different sizes from the ones used in the ATMR redundancies. In the Listing 6.1 code, for example, the ATMR uses 24-bit variables. Therefore, additional hardware will be implemented by Vivado HLS to handle the conversion from 32-bit (size of the inputs) to 24-bit variables. Each of the ATMR redundancies can be implemented with different data sizes. The data bit-sizes will affect the final result accuracy and hardware usage. Typically, if a specific data bit-size is applied to two redundancies, it will define the overall accuracy (because of the bitwise voter). However, a designer may choose different approaches to profit from the hardware cost improvement without losing precision (e.g., comparing the values considering an acceptable difference threshold and taking the output from the best accuracy redundancy as the final result). The conversions between different data sizes and types are handled by Vivado HLS. A simple cast from a different data size in the C code is enough. A more complex and probably less costly conversion could be designed, but this type of improvement is not studied. This is also the case of the ATMR voter implementation: it is left for Vivado HLS to transform the code into hardware implementation, and possible improvements are not in the scope of this book.

Six ATMR designs were implemented, varying the data precision of the operations. A non-approximate TMR version is also presented (with the three modules using 32-bit data). The designs are named following the data precision of each redundancy module to simplify the analysis and data presentation. For example, the ATMR design called "32-24-16" is composed of a module with 32-bit, one with a 24-bit and another with 16-bit precision data and operations. Those ATMR designs were applied to two matrix multiplication algorithms, one with matrices of size 3×3 and the other of size 2×2.

6.4.1.1 Accuracy Assessment

Figure 6.3 presents the inaccuracy generated by the use of approximation for each ATMR applied to the matrix multiplication operation. The data is shown in percentages and log scale. The inaccuracy value is obtained by comparing the output values of the ATMR with the one that gives better accuracy (which is the 32-bit data size multiplication due to its higher bit-size). From Fig. 6.3, it is clear that the use of fewer representation bits impacts the accuracy. As expected, if a data bit-size is applied to two different ATMR modules, it determines the inaccuracy. This is due to the ATMR voter applied to the output, which ends up considering the results from this data precision as the final one—because of that behavior; using a 24-24-24 ATMR design results in the same output accuracy as a 32-24-24 one, but

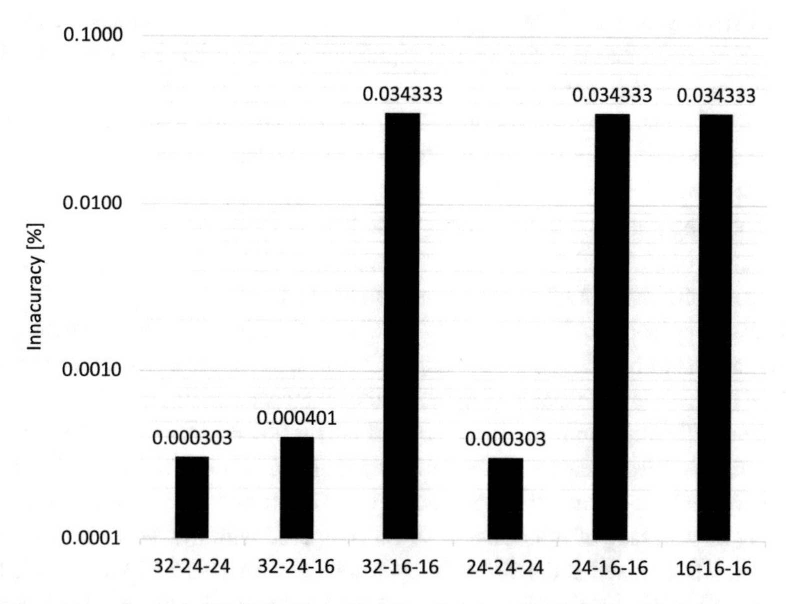

Fig. 6.3 Inaccuracy for each ATMR by data precision design applied to a 2×2 matrix multiplication. *Source* Author

with lower area usage. Another interesting outcome is the inaccuracy data for the 32-24-16 ATMR design. In this case, the inaccuracy seems to hover between the ones from the three modules.

The inaccuracy, however, is usually not high. Even in the worst case, the inaccuracy is less than 0.04%, which means the result is more than 99.96% correct. However, the increase in inaccuracy from one design to another may be a warning for more complex systems. If the inaccuracy for a complex system applying the proposed method would be significant for a 32-24-24 ATMR case, it could be considered unacceptable for the 32-16-16 one (or any situation with two modules employing 16-bit data). The ATMR variants presenting two 16-bit-size modules are two orders of magnitude more inaccurate. The inaccuracy of the 3×3 matrix multiplication design follows the same trend observed for the 2×2 matrix multiplication and therefore is not presented.

6.4.1.2 Area Usage Assessment

Table 6.3 presents the FPGA area consumption of each ATMR design for the 2×2 and 3×3 matrix multiplication operation. Data shows that approximation saves DSP usage. This behavior was already expected, given the results from Fig. 6.1. Nevertheless, the FF occupation was not predicted. The FF usage can be explained by the needed converters between the matrix multiplication operations and the voter function. The LUT area follows almost the same trend as the DSPs, decreasing with the precision reduction. The variation in the LUT usage can also be explained by the needed converters. The DSP usage for the 32-32-32 TMR design was considerably high, taking into consideration that the FPGA used in this work contains 220 DSPs. This fact highlights the importance of the approximation method presented in this work.

The first 3×3 matrix multiplication TMR design is bold to highlight the number of DSPs used. The FPGA contains 220 DSPs, while the 32-32-32 TMR design for the 3×3 matrix multiplication would require 324. Therefore, this design could not be implemented on this hardware, needing a more expensive one. With the proposed approximation, however, the implementation of an ATMR-protected 3×3 matrix multiplication is now possible. All the 3×3 matrix multiplication ATMR designs fit in the FPGA.

Comparing the data from Fig. 6.3 and Table 6.3, it is clear how the data precision reduction method is capable of reducing the area usage of the design with low effect on accuracy. The 32-24-16 ATMR design is capable of reducing the DSP usage to almost half of the 32-32-32 TMR design while introducing an inaccuracy of only 0.0004%. Another excellent example of the proposed approximation method efficiency is the results for the 16-16-16 ATMR design. It was able to reduce the DSP usage to a fourth and the FF usage in half while maintaining an accuracy of more than 99.96% compared with the 32-32-32 design. From Sect. 6.4.1.1, it is known that the 32-16-16, 24-16-16, and 16-16-16 ATMR designs have all the same accuracy. However, it is clear from Table 6.3 that the 16-16-16 ATMR design is a better choice not only because of the area usage but also due to its lower latency.

Table 6.3 Area usage and performance latency of the ATMR by data reduction designs for 2 × 2 and 3 × 3 matrix multiplications

Benchmarks		Area			Max latency
TMR design	Matrices size	DSP48E	FF	LUT	Target clock: 10 ns
32-32-32	2×2	96	1985	888	9
	3×3	**324(*)**	**7560**	**3541**	**15**
32-24-24	2×2	64	1859	761	9
	3×3	216	6543	2964	14
32-24-16	2×2	56	1763	595	9
	3×3	189	5735	2138	14
32-16-16	2×2	48	1759	945	9
	3×3	162	4576	1368	14
24-24-24	2×2	48	1815	1609	8
	3×3	162	5649	2673	12
24-16-16	2×2	32	1841	1305	6
	3×3	108	3653	1165	11
16-16-16	2×2	24	1032	689	6
	3×3	81	2257	346	9

6.4.2 Software ATMR Based on Successive Approximation and Loop-Perforation

The unique behavior of successive approximation algorithms arises as an opportunity to improve traditional redundancy fault tolerance methods. The number of iterations of a successive approximation algorithm impacts not only the accuracy of the output but also its execution time. When applying a TMR method to a successive approximation algorithm, there is no need to have three tasks with high accuracy. Because only one of the outputs will be taken as the final "correct" one, the others can have a lower accuracy (i.e., fewer iterations). Tasks with lower accuracy and execution time cause less overhead. In other words, we can naturally apply loop-perforation to a successive approximation algorithm by simply relaxing its precision.

Figure 6.4 presents the proposed ATMR method. In the figure, R1' and R2' are redundant tasks of R0 with fewer iterations, while R1 and R2 are hard copies of R0. The overhead of a TMR consists of the extra execution time it costs. Unfortunately, the overhead of the checker (represented in the figure by the CKR box) is constant. However, by reducing the execution time of the tasks, the overhead of the TMR can be lowered. Because R1' and R2' execute faster than R1 and R2, the ATMR presents a speedup in relation to the TMR.

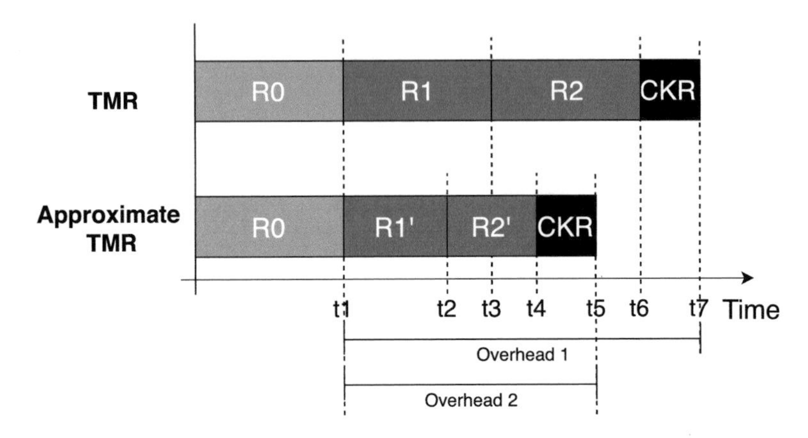

Fig. 6.4 Diagram of the proposed ATMR method. *Source* Author

Table 6.4 Execution time overheads of ATMR configurations applied to the Newton-Raphson algorithm

ATMR configuration	Execution time overhead (factor)	Execution time [ms]
71-71-71	3.09	963.268
71-71-37	2.48	771.479
71-71-14	2.22	690.381
71-37-37	1.86	579.986
71-37-14	1.60	496.237
71-14-14	1.33	414.201

Table 6.4 presents five different ATMR configurations applied to the Newton-Raphson algorithm (which will be detailed further in Chap. 7) running in a single ARM Cortex-A9 processor with data cache enabled. This algorithm is an excellent example of successive approximation used to calculate the roots of a function and will be further better explained in Chap. 7. The execution time overhead is presented in the table as a factor and is calculated in relation to a single execution of the Newton-Raphson algorithm with 71 iterations. The execution time in the last column is the total execution time of that ATMR configuration. The benchmarks are named following the number of iterations of each ATMR task (N_0-N_1-N_2, being N_n the number of iterations of the n-th ATMR task). For example, the ATMR configuration called "71-37-14" is composed of one task with 71 iterations, one with 37 iterations and another with 14 iterations. Each ATMR task may have a different number of iterations, but the algorithm remains the same. The number of iterations of each task differs because they start at different starting points and have different stop conditions. As

the table shows, the configurations with tasks that contain fewer iterations presented a lower execution time overhead.

The checker plays a critical role in the ATMR method. In a traditional TMR, the checker would make a bitwise comparison between the three outputs, changing every bit that is different from the other two to the same value. However, with approximate computing, the checker needs to be more complex. The value of the three outputs may be different even in the absence of errors, because of the varying accuracy of each ATMR task. To deal with this issue, the ATMR checker is programmed to generate as system output a midterm between the three output values of the redundant tasks. Also, when no error is present, the output from the ATMR task of better accuracy (i.e., more iterations) can be used as output, thus implying no accuracy loss. Because of this approximate checker, we have to consider a threshold of acceptable difference between the ATMR output value and the expected golden value. If the output value differs from the golden value inside this threshold limit, the ATMR is considered to have masked the error. This acceptance threshold might be different for each application or system and impacts the ATMR error masking performance. Chapter 7 will present, in Sect. 7.4, the fault masking results for the ATMR for three different thresholds: $\approx$0%, 2%, and 5%.

Another way of providing approximate computing in software is through variable data size reduction. However, using data precision reduction on embedded software is not the same as previously presented in Sects. 6.2.1 and 6.4.1. In those sections, C code is used just as a tool to generate hardware implementations with HLS. When working with embedded software, data precision reduction is more limited, as software variables are subject to predefined types. Programming new data types in software is possible, but implies a large execution overhead, given that all the operations that would otherwise be native to the hardware in use now have to be software-processed. In Sect. 7.4, where the results from this ATMR are presented and discussed, two versions of the benchmark applications will be presented, making use of *float* (32-bit, single-precision) and *double* (64-bit, double-precision) variables. Because those two types of variables are capable of achieving different accuracies, they are expected to influence the behavior of the successive approximation method. Changing the variable type for a more precise one can, for example, reduce the accuracy difference between more and less precise ATMR tasks, or make the successive approximation algorithm converge faster.

6.4.3 Parallel Software ATMR Based on Function Skipping

In Chap. 5, TMR has been used as a means to memory fault masking, protecting the system against possible SDCs. When implementing TMR with Pthreads on a modern operating system running on a multicore architecture, it can be used to tolerate other types of errors. Figure 5.7 showed that multicore applications running on Linux are especially subject to unexpected terminations. Those are errors on which the application is unexpectedly terminated by the OS due to an internal problem, such as invalid memory accesses or segmentation faults. Errors as such are also present in bare metal applications but are not manifested in the same way due to the absence of an operating system that could catch them.

As expected, the parallel versions of TMR implied a higher occurrence of UTs (Fig. 5.7). In an ordinary Linux system application, the entire application is terminated in the case of a UT on any of its threads. The problem with this approach is that it kills processes that are probably healthy and capable of providing correct data.

To deal with that problem, this section proposes using an approximation strategy based on function skipping to improve the parallel TMR proposals. This approximation strategy provides the possibility to tolerate UTs. Since every redundant TMR thread is executing the same computation, the application can finish execution even if one or two of them terminates unexpectedly. A segmentation fault handler is implemented into the TMR code to catch exceptions that would provoke unexpected terminations and, instead of terminating the application, kill only the faulty thread. We call this method "thread disposability". The implementation of thread disposability on the full sequential benchmark is not possible due to its absence of parallel threads (apart from the main one).

Figure 6.5 presents the results for the parallel TMR implementations with thread disposability. It is noticeable that the occurrence of UTs drops dramatically. In contrast, the presence of other errors remains almost unchanged. It indicates that the thread disposability technique has little influence on the normal execution of ordinary TMR. As expected, the simulation results show that different parallelization approaches imply various suscep-

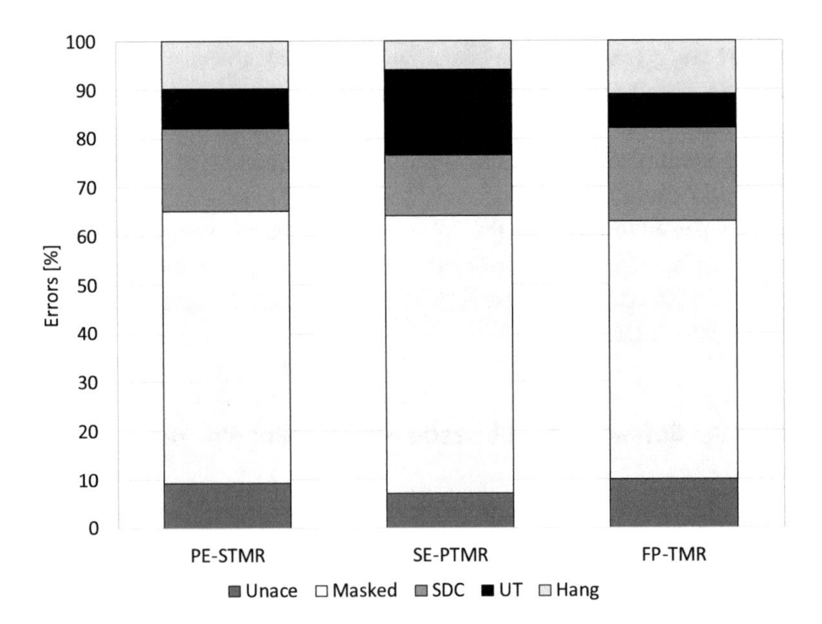

Fig. 6.5 Parallel TMR implementations with thread disposability. *Source* Author

tibilities to errors. The usage of parallelism highly increases the occurrence of unexpected terminations, that is, errors that affect the normal behavior of the application and are caught by the OS before causing severe problems. Because of that, the use of full sequential TMR appeared to be the best choice for software protection among ordinary alternatives. Using parallel TMR techniques without thread disposability causes more errors than it is capable of masking, making it unsuitable.

Thread disposability proves to be a useful method to deal with unexpected termination errors. Due to its non-intrusive nature, it costs virtually no extra execution overhead. With the use of this approach, the parallel implementations of TMR achieved similar or better results than their sequential counterpart. Thread disposability makes it possible for a full parallel TMR to have better fault tolerance than a full sequential one. It makes the use of parallel TMR techniques viable, achieving both good fault masking and overhead reduction.

This is an excellent example of how approximate computing can be applied to fault tolerance algorithms to improve them. Thread disposability not only improves a traditional fault tolerance technique by expanding its fault coverage capability but also improves its execution performance by freeing computational resources that were being wasted on useless computation. It is very hard, though, to evaluate its impact on the execution time performance, once the occurrence of errors is not predictable. However, the good results from this first experience of approximate computing on parallel embedded software is a motivation for the proposed works presented in this chapter. Apart from presenting approximation methods and studying their reliability, this chapter also proposes the usage of approximate computing as a means to improve fault tolerance.

6.5 Parallel Approximate Error Detection (PAED)

Real-time systems deal with the concept of *data freshness*, which considers that data received by the system has a specific expiration time. For example, an airplane system is continuously receiving data from the sensors. Data received one minute ago may not be valid (or even useful) anymore (e.g., the outside air temperature or a radar system that checks the distance from the ground). Those data are nevertheless critical, and a wrong value may affect the whole system's safety.

The problem when dealing with this type of data is that error correction becomes problematic. First, because of the intrinsic nature of the data, the error is often caused by a malfunction or approximation of a system sensor rather than a faulty code execution. Secondly, because of data freshness, the data shall be valid for its whole time window, which can be very strict. Therefore, data correction procedures require sophisticated implementation. The classical answer for that is the use of sanity checkers. Whenever data is not valid anymore, a flag warns the system that the current given data value cannot be trusted. For example, on avionics systems, a message might alert the pilot that current sensor data is not assured of representing the real scenario for the next couple of seconds. Such verification is usually provided through redundancy: a copy of the task that manipulates or generates the critical data is executed, and both output values are compared to check for their consistency.

As already discussed in Chap. 5, the trend of the microprocessors industry has been to move to multicore to achieve better execution performance. Safety-critical systems can use extra processing cores to improve reliability. Real-time systems can benefit from parallelization to guarantee the respect of their strict scheduled execution deadlines. The use of redundancy-based error detection techniques such as DWC [4] is more attractive on chip multiprocessor (CMP) architectures [6]. Those techniques impose a high execution time overhead on single-core processors. However, on CMP architectures, they can exploit vacant computing resources to execute redundant tasks without compromising the performance of the system.

This section proposes an error detection technique conceived for multicore real-time systems that profit from CMP architectures while following the general requirements of most real-time systems. It is designed with aerospace systems in mind and uses approximate computing to reduce the execution time overhead caused by redundant code. The new technique, called parallel approximate error detection (PAED), is adapted for multicore processors. It consists of using a processor core to execute bare metal approximate versions of tasks that are executed on the main system. The main processor runs tasks on top of FreeRTOS. The technique can be applied to any multi- or many-core processing system.

Figure 6.6 shows a graphical representation of the PAED technique. CPU0 executes the critical application to be duplicated, here represented by $task_0$, on top of FreeRTOS. CPU1 is used to execute the redundant task ($task_1$), and a checker function performs the error detection. CPU1 and $task_1$ are highly attached to CPU0 execution through dedicated synchronization primitives. Therefore, there is no need for CPU1 to execute on top of FreeRTOS too. Using bare metal to provide error detection is more reliable, as is further discussed in this book. The purpose of CPU1 is not to correct an eventual error, but to warn the system that specific data cannot be trusted. In a multicore processing system, this warning could then be used as a flag to start an error correction method.

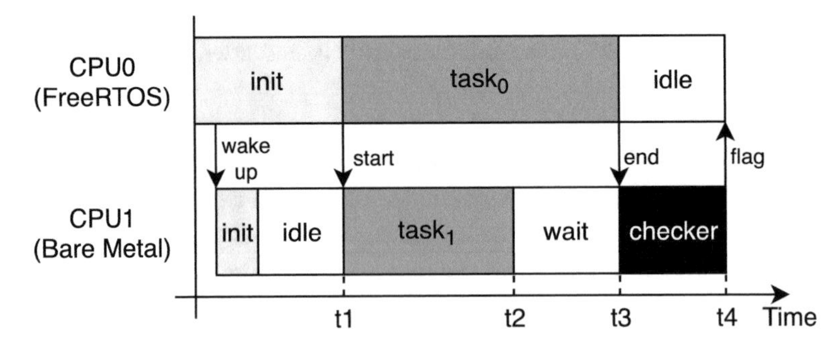

Fig. 6.6 Functional flow of the PAED technique. *Source* Author

As shown in Fig. 6.6, CPU1 is wakened up by CPU0 at the system initiation. It then waits until $task_0$ starts its execution on CPU0. The $task_1$ block is a redundant copy of $task_0$. In the figure, $task_1$ has a smaller size than $task_0$ to represent the fact that it is not an identical copy, but rather an approximate version of $task_0$. At the end of $task_0$ and $task_1$ execution (t3), CPU1 checks the integrity of the data by comparing the outputs of the tasks. If the data is not valid, a warning flag is raised. All the data management is done by CPU0, including the initialization of the inputs. The CPUs share data memory space regarding the inputs of the tasks, and all the communication flags used for synchronization. Those flags are always either read-only or write-only by each of the CPUs, avoiding memory sharing errors and race conditions. The memory space of $task_0$ is accessed by CPU1 only when the checker is executed, to verify its output value.

As previously discussed, due to the approximation, $task_0$ and $task_1$ will generate different outputs even in the absence of a fault. The accuracy loss due to the approximation has to be taken into account by the error checker, who shall accept a maximum difference threshold between the two values when comparing those. This threshold varies according to the $task_1$ approximation degree. If the values differ inside this difference threshold, the approximate checker cannot know if it was caused by a fault or by the approximation itself. A system designer shall then carefully analyze this threshold before making use of an approximate checker: some systems might not tolerate even the slightest accuracy deviation.

The idle and wait blocks in Fig. 6.6 can be used to improve system performance. In a more complex system, executing a multitude of tasks, processors could use those extra time windows to execute other functions. Even a hard real-time system could profit from it, as long as t1, t2 t3, and t4 are well defined. A real-time system designer would know the values of the time intervals defined in the figure, thus knowing with precision how much spare time CPU1 has to execute other tasks. He could then configure the system for optimal task scheduling.

When dealing with real-time systems, execution deadlines have to be taken seriously. Because of that, the error detection system works on tight synchronization. CPU1 only executes its functions when CPU0 activates the right communication flags. The processors communicate via shared memory resources, highly synchronized. The memory space of each core is well defined. Even though they share the same on-chip memory, the address range accessible to each is different. The only memory shared between the two cores are the flags and the memory that holds the input of the benchmark applications. The checker task, always executed by CPU1, is the only time when CPU1 accesses the CPU0 memory space, only to compare $task_0$ with the one from $task_1$.

6.6 Conclusion

A careful reader must have realized that the methods presented in that section are not globally applicable, i.e., they are a better fit for certain types of algorithms than others. Loop-perforation, for example, only brings optimal results when applied to algorithms that rely heavily on loops. As seen in Sect. 6.2, approximate computing in practice can significantly reduce the computational costs of numerical algorithms. However, data precision reduction methods are much more easily applied to any type of code, algorithm, or hardware design. Finally, an approximation can be naturally fault-tolerant. As discussed in Sect. 6.2.2, successive approximation algorithms have an auto-correction behavior that might make them more reliable.

Ironically, implementing fault tolerance techniques such as TMR increases the probability of a system being subject to a fault. In hardware projects, a larger design means more area subject to radiation-induced faults, for example. Similarly, in a software implementation, the more memory a system uses, the bigger is its sensitive area; with a higher execution time, the probability of a fault also increases. Thus, using approximation methods in fault tolerance techniques might make them less costly and safer.

In the next chapter, we will analyze the impacts of approximation on fault tolerance. Chapter 7 will present an evaluation of approximate algorithms under fault injections following the methodologies defined in Chap. 4, making the motivation on approximation clearer.

References

1. I. Albandes, A. Serrano-Cases, M. Martins, A. Martínez-Álvarez, S. Cuenca-Asensi, F.L. Kastensmidt, Design of approximate-TMR using approximate library and heuristic approaches. Microelectron. Reliab. **88–90**, 898–902 (2018); 29th European Symposium on Reliability of Electron Devices, Failure Physics and Analysis (ESREF 2018)
2. T. Arifeen, A.S. Hassan, H. Moradian, J.A. Lee, Probing approximate TMR in error resilient applications for better design tradeoffs, in *2016 Euromicro Conference on Digital System Design (DSD)* (2016), pp. 637–640
3. B. Barrois, O. Sentieys, D. Menard, The hidden cost of functional approximation against careful data sizing—a case study, in *Design, Automation Test in Europe Conference Exhibition (DATE), 2017* (2017), pp. 181–186
4. P. Cheynet, B. Nicolescu, R. Velazco, M. Rebaudengo, M. Sonza Reorda, M. Violante, Experimentally evaluating an automatic approach for generating safety-critical software with respect to transient errors. IEEE Trans. Nucl. Sci. **47**(6), 2231–2236 (2000)
5. W.H. Foy, Position-location solutions by Taylor-series estimation. IEEE Trans. Aerosp. Electron. Syst. **AES-12**(2), 187–194 (1976)
6. D. Gizopoulos, M. Psarakis, S.V. Adve, P. Ramachandran, S.K.S. Hari, D. Sorin, A. Meixner, A. Biswas, X. Vera, Architectures for online error detection and recovery in multicore processors, in *2011 Design, Automation Test in Europe* (2011), pp. 1–6
7. I.A.C. Gomes, M. Martins, F.L. Kastensmidt, A. Reis, R. Ribas, S.P. Novalès, Methodology for achieving best trade-off of area and fault masking coverage in ATMR, in *2014 15th Latin American Test Workshop - LATW* (2014), pp. 1–6
8. I.A.C. Gomes, M.G.A. Martins, A.I. Reis, F.L. Kastensmidt, Exploring the use of approximate TMR to mask transient faults in logic with low area overhead. Microelectron. Reliab. **55**(9), 2072–2076 (2015); Proceedings of the 26th European Symposium on Reliability of Electron Devices, Failure Physics and Analysis

9. T. Moller, R. Machiraju, K. Mueller, R. Yagel, Evaluation and design of filters using a Taylor series expansion. IEEE Trans. Vis. Comput. Graph. **3**(2), 184–199 (1997)
10. H. Quinn, Challenges in testing complex systems. IEEE Trans. Nucl. Sci. **61**(2), 766–786 (2014)
11. G.A. Reis, J. Chang, N. Vachharajani, S.S. Mukherjee, R. Rangan, D.I. August, Design and evaluation of hybrid fault-detection systems, in *32nd International Symposium on Computer Architecture (ISCA'05)* (2005), pp. 148–159
12. Y. Ren, B. Zhang, H. Qiao, A simple Taylor-series expansion method for a class of second kind integral equations. J. Comput. Appl. Math. **110**(1), 15–24 (1999)

Experimental Analysis and Discussion 7

7.1 Summary

Throughout this book, we have been discussing approximate computing, fault tolerance, and their relation with safety-critical systems. Chapter 6 presented approximated computing algorithms and approximated versions of traditional fault tolerance techniques. Although that chapter showed some intuitions and data regarding the cost reduction and execution time performance improvement of approximation, it did not present real proofs of the good and reliable approximation behavior in real case scenarios. This chapter, in contrast, will focus on a more profound discussion over their computational costs, testing those methods and techniques under simulations of real-case safety-critical circumstances and evaluating them for their performance under fault injection scenarios.

The testing and evaluation methodologies defined in Chap. 4 will be put in practice here. Notice that not all methodologies defined in Chap. 4 are applied to all approximation and fault tolerance techniques, because each of them has its most appropriate evaluation methodology. A sufficient number of errors was gathered from each of the experiments to obtain statistically significant results with an error margin of 1% and a confidence level of 95% using the approach presented in [2].

The chapter is structured as follows. Section 7.2 presents discussions on the proposed approximation methods implementations, their implementation costs on hardware and software, and results regarding their reliability under emulation and laser fault injections. The Taylor series approximation is the only proposed technique which will be not tested under fault injection experiments. Instead, Sect. 7.2.1 will analyze its implementation costs and impact on the system accuracy of hardware and software. All other proposed works will be evaluated both for their fault tolerance and computational resource cost (e.g., execution time and programmable hardware area). Sections 7.3 and 7.4 present the results for the ATMR

G. S. Rodrigues et al., *Approximate Computing and its Impact on Accuracy, Reliability and Fault-Tolerance*, Synthesis Lectures on Engineering, Science, and Technology,
https://doi.org/10.1007/978-3-031-15717-2_7

behavior under onboard and laser fault injections. Finally, Sect. 7.5 presents the results for the approximate error detection technique under laser fault injection.

7.2 Approximation Methods

This section starts by presenting a discussion regarding the implementation of Taylor series approximation on hardware and software, their development, implementation costs, and impact on accuracy in Sect. 7.2.1. Then Sect. 7.2.2 presents a study of the successive approximation under emulation and laser fault injection. Finally, Sect. 7.2.3 presents a study on the behavior of approximate algorithms compared to non-approximate traditional applications, executing bare metal and on top of operating systems (FreeRTOS and Linux).

7.2.1 Taylor Series Approximation

The data from Tables 6.1 and 6.2, presented and discussed in Sect. 6.2.3, arise a multitude of questions. They show that the hardware and software implementations have their particularities. A comparison between the area resource usage from HLS and embedded software does not make sense, because the area of the ARM processor is constant and not dependable on the program implementation. However, it is possible to compare some data from hardware and software. As an example, the latency of HLS hardware implementations (particularly the ones without pipeline) is comparable with the execution latency of the embedded software implementation. This section discusses the results obtained from Sect. 6.2.3 and speculates on their implications.

7.2.1.1 Hardware Implementation Analysis

Section 6.2.3 presented the Vivado HLS implementation details, in Table 6.1. The usage of pipeline arises as a good alternative for projects that need to rely on fast execution, and which are implemented on boards on which area is not a problem. In fact, using pipeline may even not be costly. For instance, comparing the pipelined version of double-precision with its counterpart with no pipeline, Table 6.1 shows that when both achieve an accuracy of more than 99% the pipelined version uses around four times the number of LUTs and FFs but is almost seven times faster. Similar behavior is observed on the *float* variants (also called single-precision, i.e., 32-bit floating-point format).

Hardware area resource usage can be a problem for some systems. As was explained in Sect. 6.2.3, the maximum number of Taylor series terms that fit in the Zynq board FPGA are 13 and 34 for double- and single-precision, respectively, when making use of pipeline. Nevertheless, every benchmark variation shows that the accuracy of the Taylor series approximation largely increases following a small number of terms. Because of that, the need for a high number of terms is improbable. Projects in need of high accuracy can also achieve it while saving area by limiting the size of the pipeline, and breaking the computation loop into big chunks. This type of implementation strategy is not analyzed in this work. Table 6.1 also shows that increasing the area of the design increases the number of essential bits. This can be seen as problematic for safety-critical systems since it is highly related to the project susceptibility to errors when exposed to radiation [5]. As referenced in Sect. 6.2.3, the type of approximation presented in this work might improve the fault tolerance of an application. If that is the purpose of the approximation, a designer may choose not to use the pipeline approach.

The accuracy (defined by the number of Taylor series terms) is determined by different factors for each variant of the benchmark. The versions without pipeline implementations have their accuracy determined by their area size. However, the pipelined versions achieve accuracy by increasing their latency. A designer shall know where he shall pay the cost of accuracy: whether in latency or area. The advantage of using implementations with no pipeline is that, despite taking a long time to output, it is always capable of achieving the best accuracy possible. That is because time is a resource not limited by the hardware, but by project constraints. The same cannot be said of pipelined implementations: even though they are faster than the designs that do not use them, they have their maximum accuracy limited by the programmable area of the hardware.

Table 6.1 shows that improving the accuracy of an already accurate version of the algorithm is more costly than enhancing an inaccurate one. As a good example, there is the double-precision pipelined version. In that case, improving the accuracy from 54.7% to 89.28% is less costly in the area than improving it from 99.994% to 99.999%. The same behavior is observed in all versions of the code. It shows that the higher the index of the Taylor series term, the lower is its impact on the final result, and thus less critical it is. It also indicates that there is a maximum accuracy attainable by the approximation method, but it is very close to 100%. The table proves the importance of a preliminary study to avoid unnecessary or unworthy area usage. For instance, there is no reason to use 100 Taylor terms on the double-precision without the pipeline version because 25 is already enough to achieve an accuracy of 100%.

7.2.1.2 Software Implementation Analysis

Section 6.2.3 presented on Table 6.2 the performance details from the Taylor series implementation running as a bare metal application on the ARM A9 processor. The double-precision variant of the algorithm was the only one capable of achieving an accuracy of 100%. Nevertheless, the single-precision met good accuracy with low execution latencies.

Contrary to the hardware HLS approach, the only cost for embedded software to improve the accuracy is execution latency. The memory use of the variants is also different, but the absolute value is so small that it has no cost impact. The output is only one 64- or 32-bit variable, for *double* and *float*, respectively. Unless a project needs accuracy of 100%, there seems to be no reason to use double-precision instead of single-precision. Nevertheless, it is important to remember that the accuracy of each algorithm variant is calculated by taking as a parameter its own best result possible (assuming it to be the *math.h* library result). It means that double-precision provides not only accuracy, but also a more precise result. The type of data precision defines how many decimal points the variable can hold. This number may change depending on the target processor architecture or compiler used, but normally double-precision holds thirteen decimal points, while single-precision holds seven. For projects that need high exactitude, the seven decimal points provided by single-precision may not be enough. In those cases, double-precision is a must. However, as Table 6.2 shows, Taylor series can provide accuracy with almost the same computation execution latency as single-precision (*float*).

7.2.1.3 Discussion on the Software and Hardware Implementations

The software implementation achieves accuracy by increasing the number of Taylor series terms, which in its turn increases the execution latency. In the HLS hardware implementations, increasing the number of terms would cause an increase in area and latency (being a higher variation in the area for the pipelined variants and higher latency for the ones without pipeline). Because the pipelined hardware increases area (and almost no latency) to achieve accuracy, it is not fair to compare it with the software implementations. The hardware implementations without pipeline, however, are comparable with the software implementations, as both achieve accuracy at the same price: latency.

Figure 7.1 presents the data from software and hardware execution latency for the double-precision algorithms. The black vertical lines mark some important precision barriers. Fol-

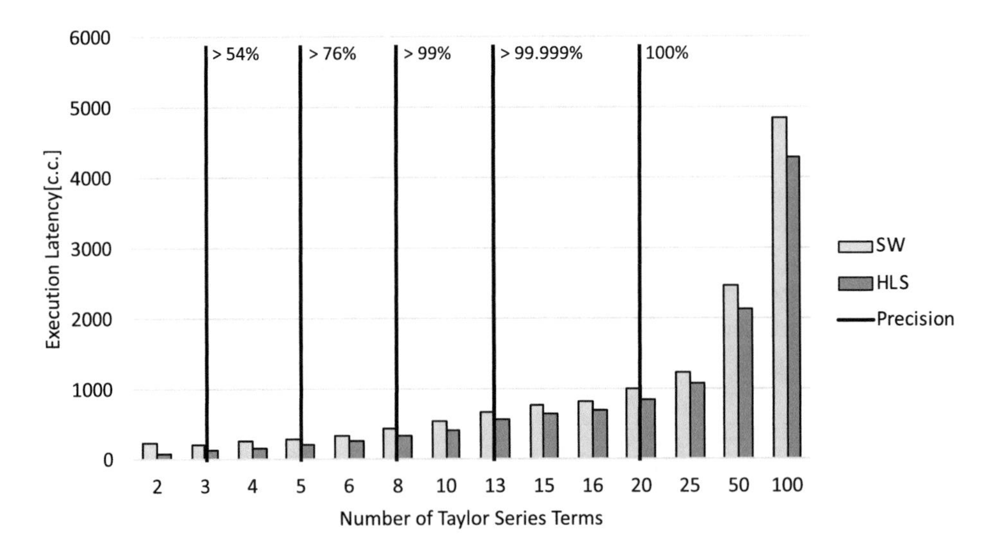

Fig. 7.1 Execution latency comparison between HLS without pipeline and SW (software) implementations for double-precision. *Source* Author

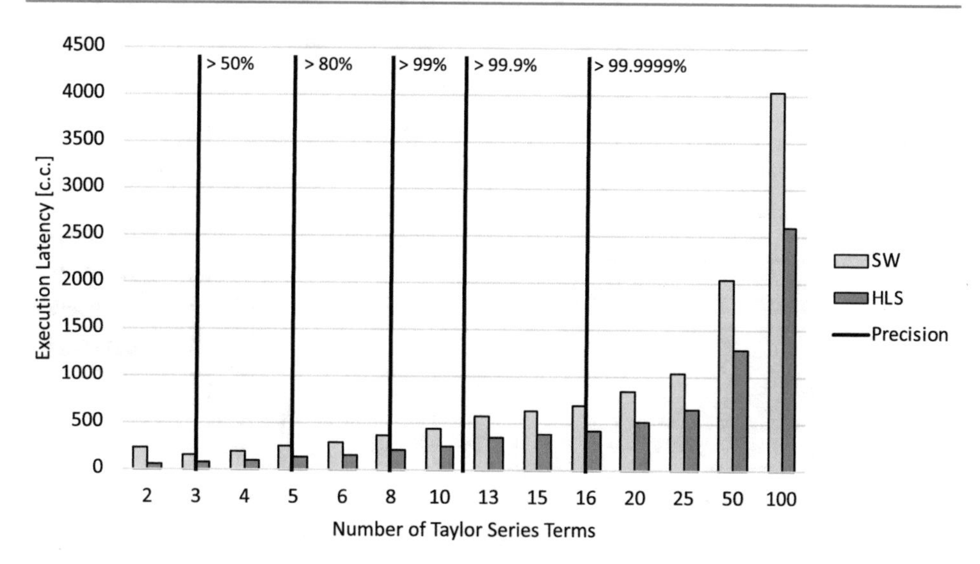

Fig. 7.2 Execution latency comparison between HLS without pipeline and SW (software) implementations for single-precision (*float* type variables). *Source* Author

lowing the tendency observed in Tables 6.1 and 6.2, the precision increases faster for the first terms, and slower as it gets near 100%. The unexpected result is that the hardware HLS and software (SW) implementations had virtually the same execution in clock cycles.

The comparison between the HLS hardware without pipeline and software implementations with single-precision regarding execution latency is presented in Fig. 7.2. In this case, the software implementation takes more clock cycles to finish than the hardware one. The line progression shows that the number of required clock cycles rises exponentially, as well as the difference between their absolute values. It indicates that if a higher number of terms were needed, the software approach would take much more clock cycles to finish the execution than the HLS.

Studying the execution performance only through the number of clock cycles may be misleading. That is because the ARM processor and the PL of the Zynq-7000 APSoC have different frequencies. The embedded ARM processor works with 666MHz, while the FPGA on the PL runs with a frequency of 100MHz. It means that a software version of the algorithm may execute faster than an HLS hardware implementation even with a higher clock cycle count. Figure 7.3 presents the execution time in seconds of the two data precision variants from HLS hardware (without pipeline) and software implementations. It shows that software implementations are always faster than hardware with no pipeline. This is an unexpected result, taking into account that hardware implementations tend to be faster than embedded software. In contrast to Figs. 7.1 and 7.2, 7.3 shows that a more general and less optimal design (with a higher clock cycles count) can execute faster than an optimal one (with fewer clock cycles). It all depends on the target hardware. It is important to notice, however, that

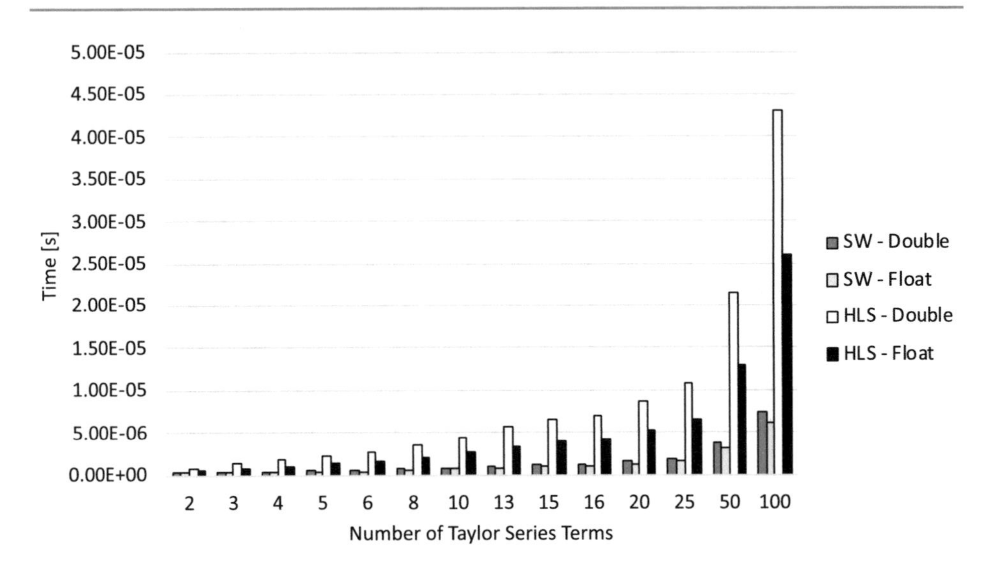

Fig. 7.3 Time comparison between HLS without pipeline and SW (software) implementations for both data precisions. *Source* Author

the hardware implementations in Fig. 7.3 do not take full profit of the parallelization capacity of the FPGA since they are not implementing pipelines and loop unrolling.

The question of whether to use hardware or embedded software to implement approximation through the Taylor series seems to have no definite answer. While software arises as a better alternative than low-area hardware, it is still slower than an optimal pipelined hardware implementation. However, the high area cost of a pipeline for the Taylor series with a high number of terms may prove its implementation unfeasible on smaller FPGAs. A profound evaluation of the alternatives shall be performed before a design decision, since project time and area constraints may vary.

7.2.2 Successive Approximation and Loop-Perforation

Three successive approximation numerical algorithms are presented and used as benchmark applications:

- **Newton-Raphson**: The Newton-Raphson method is an algorithm used to find the roots of a function. It calculates the intersection of the tangent line of the function in an initial guess point x_0 with the x-axis. It is calculated iteratively, as stated in (7.1), until it reaches a sufficient approximation.

$$x_{n+1} = x_n - \frac{f(x_n)}{f'(x_n)} \tag{7.1}$$

- **Trapezoid Rule**: As already discussed in Sect. 6.2.2, the trapezoid rule algorithm is used to calculate the integral of a function. It approximates the area under a curve to some trapezoids and then calculates their areas. Considering N equally spaced trapezoids defined between points a and b of the function, we have each trapezoid k with a base of length $\Delta x_k = \Delta x = \frac{b-a}{N}$. The integral approximation with the trapezoid rule is defined in (7.2).

$$\int_a^b f(x)\,dx \approx \frac{\Delta x}{2} \sum_{k=1}^{N} (f(x_{k-1}) + f(x_k)) \tag{7.2}$$

- **Simpson**: Another way of numerically approximating a function integral is with Simpson's rule. The difference between Simpson's rule and the Trapezoid is that it calculates the area of parabolas instead of trapezoids. This way, it usually approximates the result with more exactitude in fewer iterations than the Trapezoid rule. The Simpson approximation for an integral with a step size of $h = (b - a)/2$ is presented in (7.3).

$$\int_a^b f(x)\,dx = \frac{h}{3}\left[f(a) + 4f\left(\frac{a+b}{2}\right) + f(b)\right] \tag{7.3}$$

Each one of those benchmarks has three variants. For each variant, the number of iterations is different (and therefore the accuracy), but the algorithm remains the same. Table 7.1 provides some details on each of those variants and how the number of iterations affects

Table 7.1 Successive approximation experiment benchmark details

App.	Var.	Num. of iters.	Used registers	L1 data cache accesses [per ms]	Exec. time [ms]
Simpson	1	242	r2, r3, r11, pc, sp, lr	178.2k	0.94
	2	423		1648.1k	9.31
	3	3081		2350.4k	18.62
Trapezoid	1	128	r0, r2, r3, r11, pc, sp, lr	6053.2k	202.34
	2	1274		6792.4k	605.29
	3	12746		6763.1k	33540.09
Newton-Raphson	1	14	r2, r3, r11, pc, sp, lr	97.2k	0.44
	2	37		202.4k	1.19
	3	71		682.8k	3.19

the application execution. Some algorithms converge faster to a final acceptable result than others (notice the difference between Trapezoid and Newton-Raphson in Table 7.1), and therefore naturally present a lower number of iterations. The different iteration numbers of the benchmarks allow a more in-depth assessment of how it impacts the algorithm execution behavior in relation to reliability.

The benchmarks are tested under laser fault injection at the L1 data cache memory and fault injection emulation on the register file. The results are used to assess and discuss how the successive approximation, inherent to those algorithms, affects their fault tolerance. For that purpose, each benchmark variation is compared with each other, so that the number of iterations' impact on the fault tolerance is evaluated. Faults affecting the register file are expected to have a higher probability of vanishing than the ones affecting the cache memory. As Table 7.1 shows, the proposed benchmarks are far from using all registers available. The low use of registers means that the sensitive area of the register file is small. Therefore, faults affecting it may touch registers that are not even in use. Injecting faults in the cache memory, however, may lead to unexpected behaviors. Some of the benchmarks have a high number of cache memory accesses. It can cause the fault to be read into the applications and provoke an error or faulty memory space to be overwritten, causing the fault to vanish.

The fault tolerance of the benchmarks is evaluated in two aspects. First, we assess how the number of iterations impacts the error susceptibility, i.e., how each variant presented in Table 7.1 behaves under fault injection. Variating the number of iterations for each benchmark has a significant impact on fault tolerance. This evaluation is made with results from both laser and emulation fault injection. Figures 7.4, 7.5, and 7.6 present the error occurrence (in percentage) for each type of error. Figures 7.7, 7.8, and 7.10 present the error relative probability (per laser pulse) for each benchmark and their variants, as presented in Table 7.2. This probability is calculated by normalizing the error occurrence values with their maximum for each benchmark. The normalization is needed because the error occurrence depends on the execution time and the shots per execution of the benchmark, and those are very different for each application.

Secondly, we evaluate how tolerating small variances in the output value can reduce the number of considered SDC-type errors. For that assessment, we compare the output values with the golden value and check how different they are. So for example, if an application can tolerate an output variation of 2%, an output value will only be considered erroneous if it is less than 98% equal to the golden value. This evaluation was made from the results from the laser fault injection only.

7.2.2.1 Fault Injection Emulation on Register File

Contrary to the laser fault injection, the emulation fault injection is programmed to inject one fault per execution of the algorithm. This would be impossible on laser fault injections due to the frequency of the laser pulse and the delays of the experimental system, and because some of the benchmarks are very fast. The details of the emulation fault injection differ

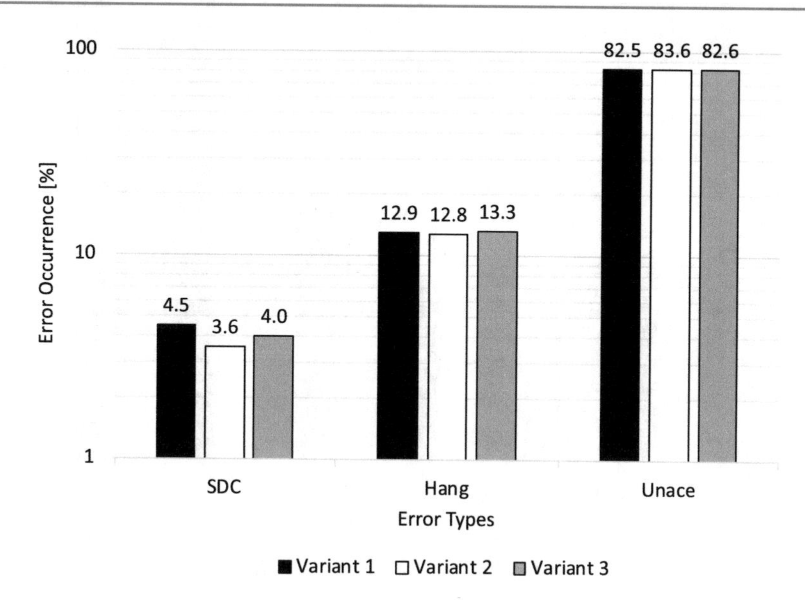

Fig. 7.4 Simpson error occurrence for emulation fault injection. *Source* Author

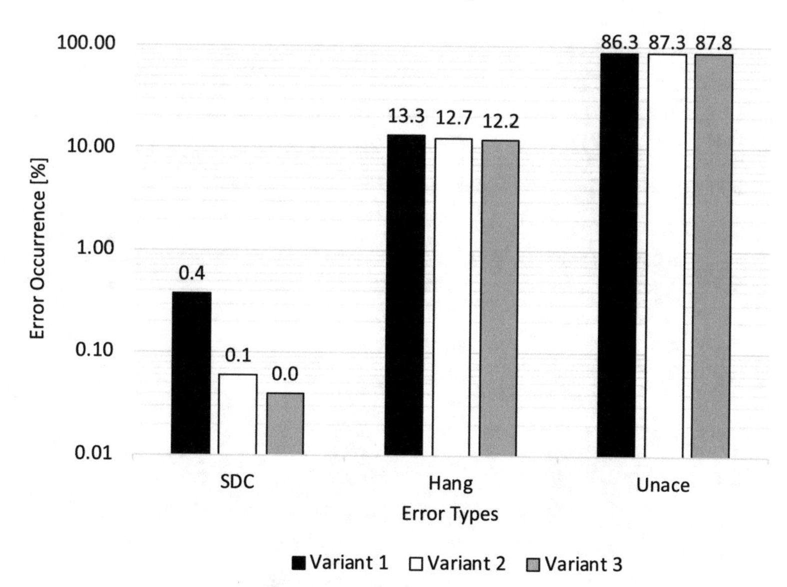

Fig. 7.5 Trapezoid error occurrence for emulation fault injection. *Source* Author

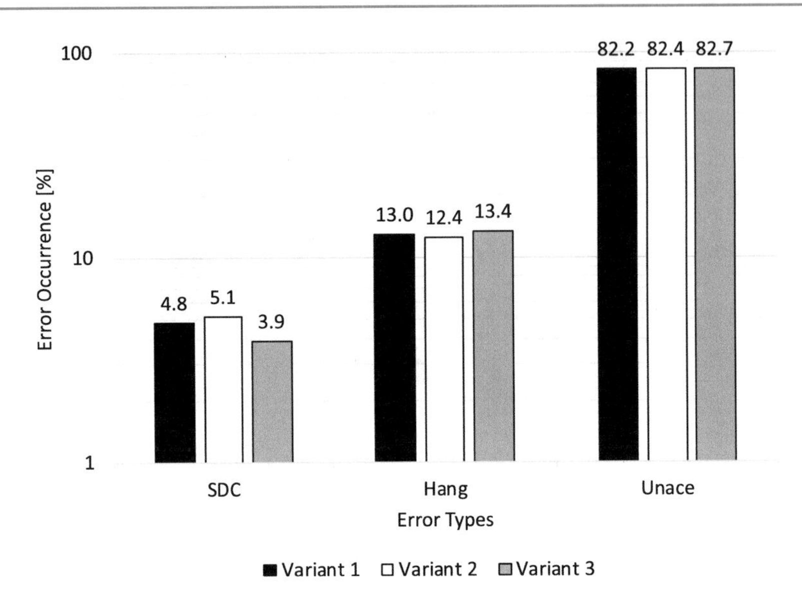

Fig. 7.6 Newton-Raphson error occurrence for emulation fault injection. *Source* Author

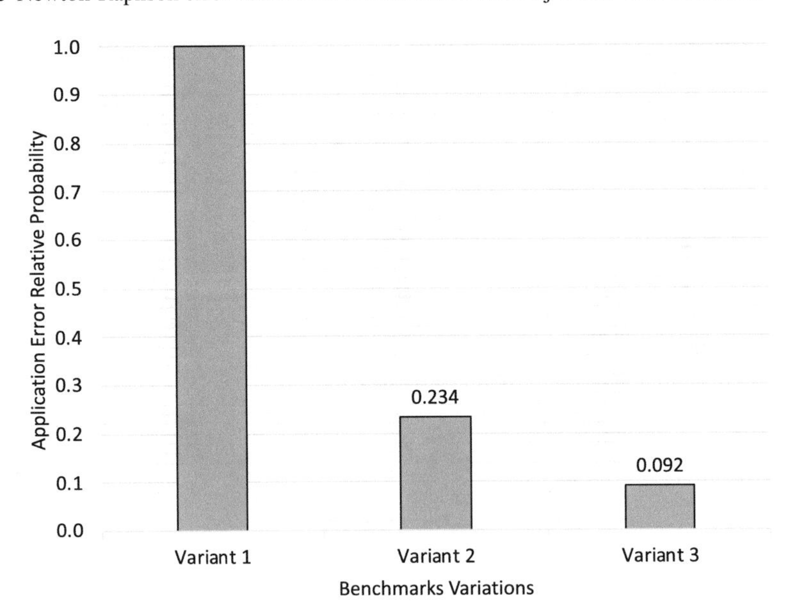

Fig. 7.7 Simpson error relative probability (per laser pulse). *Source* Author

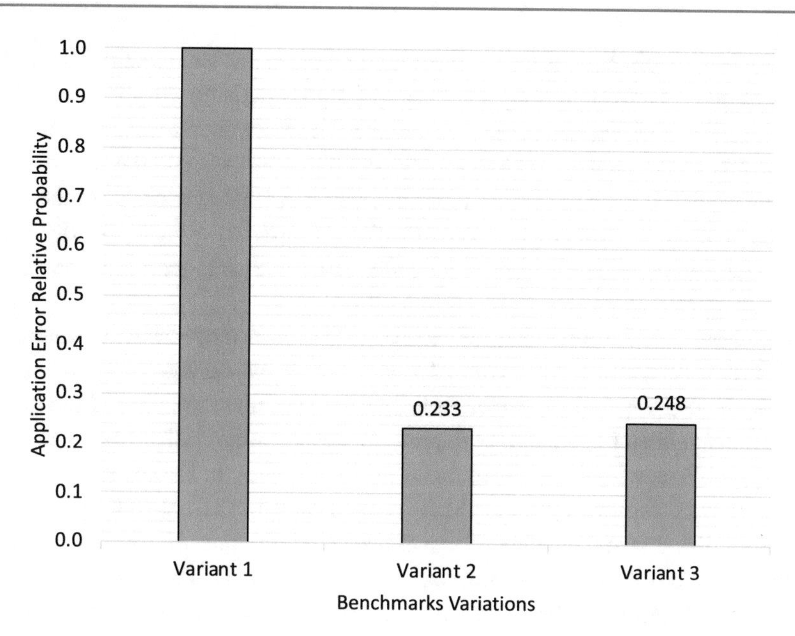

Fig. 7.8 Trapezoid error relative probability (per laser pulse). *Source* Author

Table 7.2 Laser fault injection details on successive approximation benchmarks

App.	Var.	Num. of runs (N)	Total workload [Bytes]	Avg. shots per Exec.	Exec. time [ms]
Simpson	1	100	400	0.9669	96.69
	2	100	400	9.4887	948.87
	3	100	400	18.9963	1899.63
Trapezoid	1	150	1200	302.8096	30280.96
	2	70	560	444.4170	44441.70
	3	1	8	546.1586	54615.86
Newton-Raphson	1	100	800	0.417	41.72
	2	100	800	1.302	130.20
	3	100	800	3.379	337.91

from laser ones not only due to its different characteristics but also because of the different focus of this methodology. All the interesting data on this type of injection was presented in Table 7.1. The most important information about it is the number of registers in use. The execution time is also important because it may define the probability of an error to be corrected (due to a higher amount of iterations), but the emulation fault injection assures that there will be only one fault injected per execution, no matter the execution time.

The data from Table 7.1 is gathered from the implementation at the Zynq-7000 APSoC. This was gathered concerning only the general-purpose registers (from r0 to r12 and the stack pointer, link register, and program counter). The table shows that those benchmarks tend to use few registers. It also shows that the number of accesses to the L1 Data Cache is heavily impacted by the number of iterations of the loop, with the exception of the Trapezoid application. In that case, it is probably because all the Trapezoid variants already have a high data cache access, possibly the highest possible. The fact that Trapezoid is the benchmark with the highest execution time supports that idea.

Figure 7.4 presents the percentage of each error type occurrence from the emulation fault injection at the register file. The "Unace" bars show the percentage of faults that did not cause an error. The y-axis is presented in log scale to facilitate the view of the data, given that there are significant differences between the occurrences. In that case, increasing the number of iterations of the algorithm has little to no effect on the distribution of errors. The high number of unaces shows the faults tend to vanish. As discussed before, faults are expected to vanish due to the nature of successive approximation. However, it is interesting to see that they did not vanish the same way the ones injected at the cache memory did (as will be presented further in Sect. 7.2.2.2). In that case, all the variants had the same fault tolerance.

Figure 7.5 presents the percentage of each error type occurrence from the emulation fault injection for the Trapezoid algorithm. For that algorithm, the occurrence of SDCs dropped while increasing the number of iterations. However, the SDC occurrence for variant 1 was already very low. Comparing the results from Figs. 7.5 and 7.4, it is clear that the Trapezoid algorithm is much less prone to SDCs than Simpson.

The emulation fault injection that results for the Newton-Raphson benchmarks are presented in Fig. 7.6. Again, the variation in the number of iterations did not affect the type of error distribution. Hangs are also more frequent than SDC, which is expected given that those are iteration-based algorithms. A significant part of the execution concerns loop management. Therefore, it is a definite critical point of failure. Still, most of the faults (around 82%) caused no errors.

7.2.2.2 Laser Fault Injection on Data Cache Memory

The benchmarks execute on loop under the laser fault injection, fulfilling an output vector that is later checked for SDC errors. This was made to extend the execution time of the benchmarks and assure an almost random fault injection on different points of the algorithm execution. Table 7.2 presents the details of the laser fault injection experiments, applied to each benchmark. The number of runs is the size of the output vector, i.e., the value of N times an algorithm runs per execution. The "Total Workload" represents the size in bytes of the output vector. With that data, it is possible to infer the *workload per run* (the size of the outputs) by simply dividing the number of runs N by the total workload per run. The "Execution Time" is the time of a complete execution (N runs). Finally, the "Average Shots per Execution" column presents the average number of laser shots per execution, which is calculated by dividing the execution time and the time between laser shots (i.e., the inverse value of the laser frequency). Notice that the number of runs N is different from the number of iterations presented in Table 7.1. On each run, the benchmarks execute the number of iterations defined by each variant in Table 7.1.

The application error relative probability per laser pulse of the Simpson benchmark is presented in Fig. 7.7. As expected, the variants with a larger number of iterations are more fault-tolerant. However, more iterations mean more latency. As Table 7.2 shows, variant 3 of the Simpson benchmark is almost 20 times slower than variant 1, but Fig. 7.7 shows the error occurrence does not decrease in the same pace. It means that, for this algorithm, increasing the number of iterations improves reliability, but the price is high.

The Trapezoid rule also shows a significant improvement in reliability for higher numbers of iterations, but it tends to stabilize at a certain point, as Fig. 7.8 shows. This is because the Trapezoid rule converges slower than the Simpson method. For that same reason, the number of iterations for each version of this benchmark is higher than the other ones (see Table 7.1). Variants 2 and 3 of the Trapezoid had a very similar result. It indicates that the benchmark might have an optimal point of fault tolerance on around 1200 iterations (according to Table 7.1). Using more iterations than that would add more execution time to the algorithm, but has no impact on fault tolerance. Nevertheless, as Table 7.2 shows, the execution time difference among the three Trapezoid variants is not as big as in other benchmarks.

According to Table 7.2, the execution time of Trapezoid is much longer than Simpson. Given the fact that both algorithms are applied to solve the same problem (calculating an integral), we can draw interesting conclusions by comparing both their results. Figure 7.9 presents the values of the application error relative probability per laser pulse calculated and normalized for Simpson and Trapezoid together. It is noticeable that Trapezoid is much more fault-tolerant than Simpson. This result indicates that having a higher number of iterations is beneficial for fault tolerance, but some applications might pay a high price for that. It is also clear that for this kind of approximate computing algorithm, the drawback for increasing reliability is execution time. When using an approximate computing technique to solve a computation problem, different approaches will provide very different fault tolerances, even if they are similar.

Newton-Raphson presents a behavior similar to Simpson. Figure 7.10 indicates that variant 2 of this benchmark already achieves a considerable fault tolerance improvement, having a relative probability two orders of magnitude smaller than the first variant. It is interesting to notice that this benchmark is the one with the lower number of iterations, as reported in Table 7.1. What it indicates is that the number of iterations alone is not enough to provide a fault tolerance estimation. Different from the other two, this benchmark is not used to calculate an integral, but the roots of a function. It also has a very convergent nature, so a high number of iterations is not necessary.

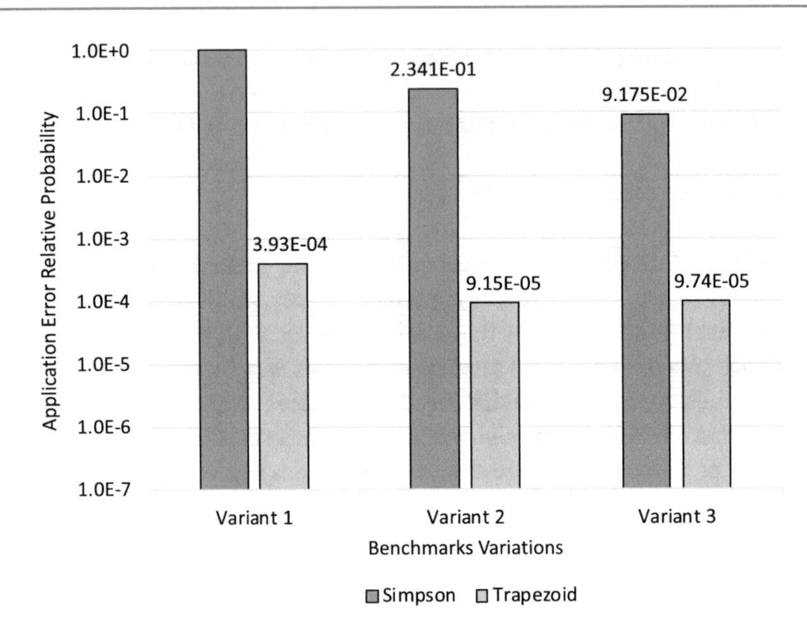

Fig. 7.9 Application error relative probability (per laser pulse) calculated for Trapezoid and Simpson together. *Source* Author

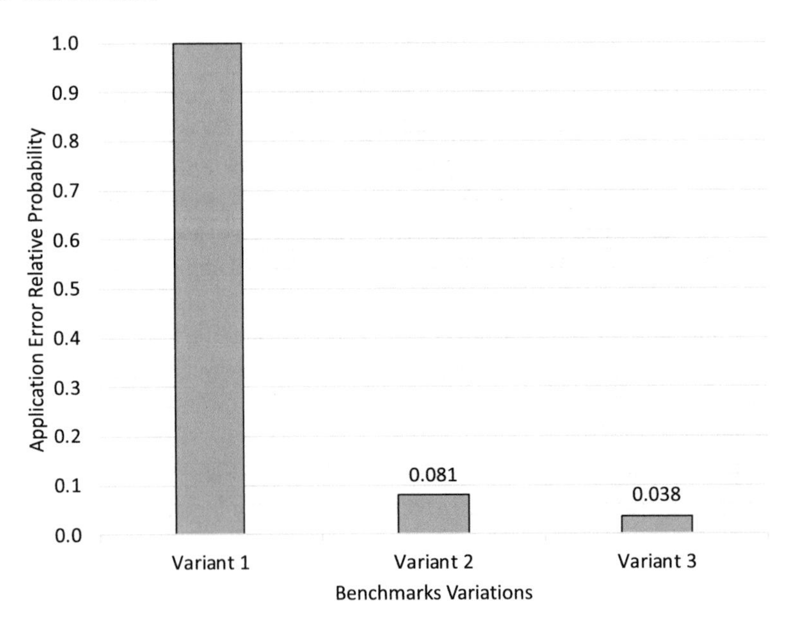

Fig. 7.10 Newton-Raphson error relative probability (per laser pulse). *Source* Author

It is interesting to see that the fault effects on the register file and cache memory are very distinct. While the faults injected at the cache tend to vanish for a higher number of iterations, the ones injected at the register file have almost the same effect no matter the loop size (the exception of the Trapezoid benchmark is noticeable, but the difference between the variants SDC error occurrence is still meager). Two facts can explain it: the register usage of the benchmarks is very low, and the data cache memory usage is crucial. As Table 7.1 shows, the benchmarks do not use all the registers. However, the fault injection on the register file is considering all of them. Thus, the probability of a fault affecting a register being used is not very high. The way registers are used also affects their criticality: they are continually being overwritten, and so are the faults injected into them. The data cache memory has a higher data latency (i.e., data usually stays untouched longer than at registers), and is also where most of the results are stored (while registers are used not to store data, but mainly to process it). Therefore, faults injected in the data cache have a more significant probability of spreading to the final output of the application.

Figure 7.11 presents the results for the error reduction when accepting output variations for the Simpson benchmark. It is clear that having an output variation tolerance of about 2.5% is enough to have a significant reduction in error occurrence. Each variant presented different results in that evaluation, but a general trend is clear. Most of the errors on this application's outputs are small, i.e., the final value does not differ much from the expected one. An approximate computing system, which is able to tolerate those small errors, may benefit from this output relaxation to provide reliability.

Figure 7.12 presents the output variation tolerance effect on the number of perceived errors for the Trapezoid algorithm. It has a very different behavior from the other benchmarks. In the worst case, for variant 2, the occurrence of SDC errors drops more than 25%, but it remains even when accepting more significant variations. The other variants presented a drop in total SDC errors of about 70% when accepting up to 2.5% output variation from the golden value. This unexpected Trapezoid behavior can be explained by its already low error occurrence. Because Trapezoid already presented much fewer errors than Simpson, it has fewer errors to tolerate. Thus, a more considerable amount of those is significant.

The error drop when variating the output value error tolerance for the Newton-Raphson algorithm is shown in Fig. 7.13. It also presents the same trend as the Simpson benchmark. Variant 3 of Newton-Raphson is the one that had a lower drop in error count while increasing tolerance. It means that the errors had a higher difference in relation to the expected output; in other words, they were "more erroneous". The error drop stagnates after around 4% of output variation tolerance.

All the results from the output tolerance variation show significant error drops. Tolerating small deviances in the expected output of an algorithm is the very definition of approximate computing. Those results indicate that approximate computing not only can be applied to safety-critical systems, but might even improve their fault tolerance. It is important, however, to notice that some systems may not tolerate even minimal output deviations. Those are not good candidates for any approximate computing technique.

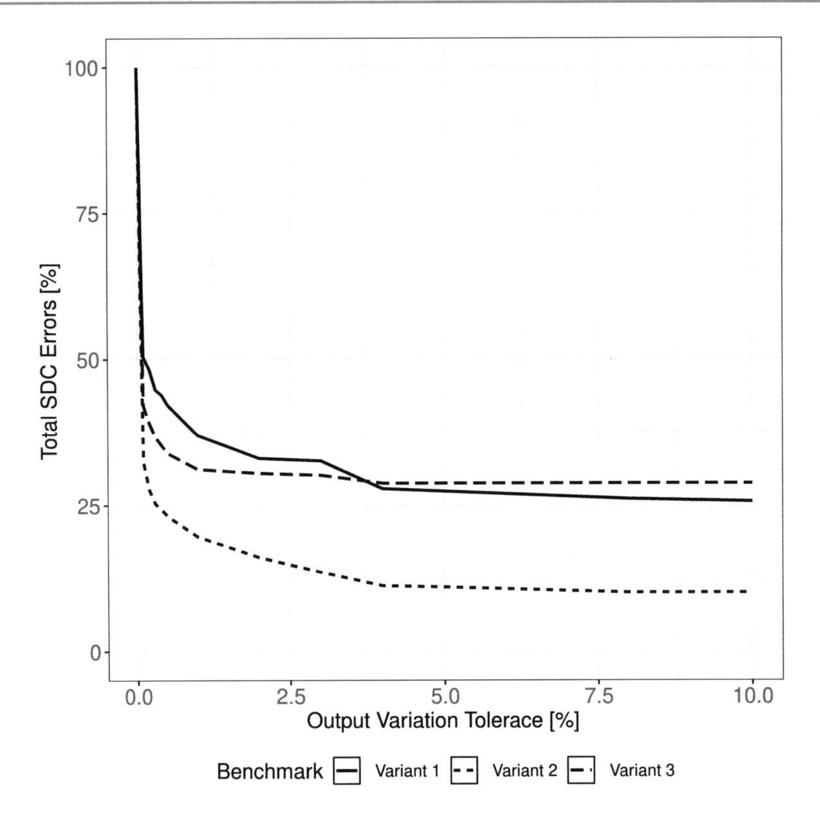

Fig. 7.11 Error occurrence drop in relation to output variation tolerance for Simpson benchmark. *Source* Author

7.2.3 Behavior and Application Evaluation on Operating Systems

Safety-critical systems may need to manage the execution of many applications, sharing resources. To guarantee the safe management of those resources, the use of an operating system is attractive once running bare metal applications on a system could lead to a waste of resources. As discussed in Chap. 5, it is important to evaluate the possibility of OS usage and its effects on the system behavior and fault tolerance. Also, knowing that successive approximation has an impact on fault tolerance, it is imperative to evaluate how it co-relates with traditional algorithms (non-approximate), as well as how it behaves when executing on top of a complex operating system.

One cannot expect that two different applications will behave similarly, even when executing in the same hardware and the same physical conditions. It is also shown that the system on which an application is executing is a significant factor for the error sensibility evaluation: some of our past works proved that an application's fault tolerance might differ a lot when executing bare metal or on top of a complex operating system such as Linux.

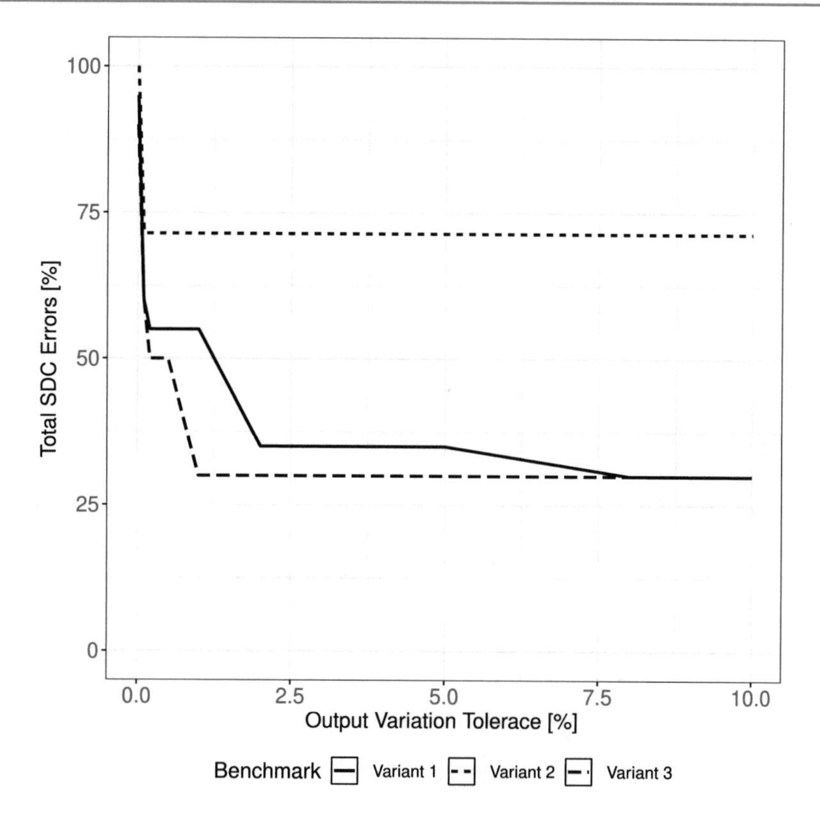

Fig. 7.12 Error occurrence drop in relation to output variation tolerance for Trapezoid benchmark. *Source* Author

In this section, we present an evaluation of successive approximation algorithms under fault injection simulation, executing both on bare metal systems and on top of FreeRTOS and Linux. We also compare their results with some ordinary computation benchmarks. Given the data from related works, it is expected that a lighter OS such as FreeRTOS will have less impact on the system susceptibility to errors than a more robust, complex one such as Linux. It is natural for a more complex system to have more critical points of failure. Therefore, we expect bare metal applications to be less susceptible to errors.

This part of the work uses the fault injection simulation implemented at OVPSim, and is presented in Sect. 4.5. The fault injection is simulated in this part because the usage of Linux OS makes a physical experiment of fault injection very difficult. Because the operating system has a boot time to initialize itself and takes a big part of the computational resources of the processor, physical fault injection methodologies do not work very well. Using OVPSim fault injection, we were able to inject faults only in the part of the execution that was interesting for this work: the application running on top of the operating system.

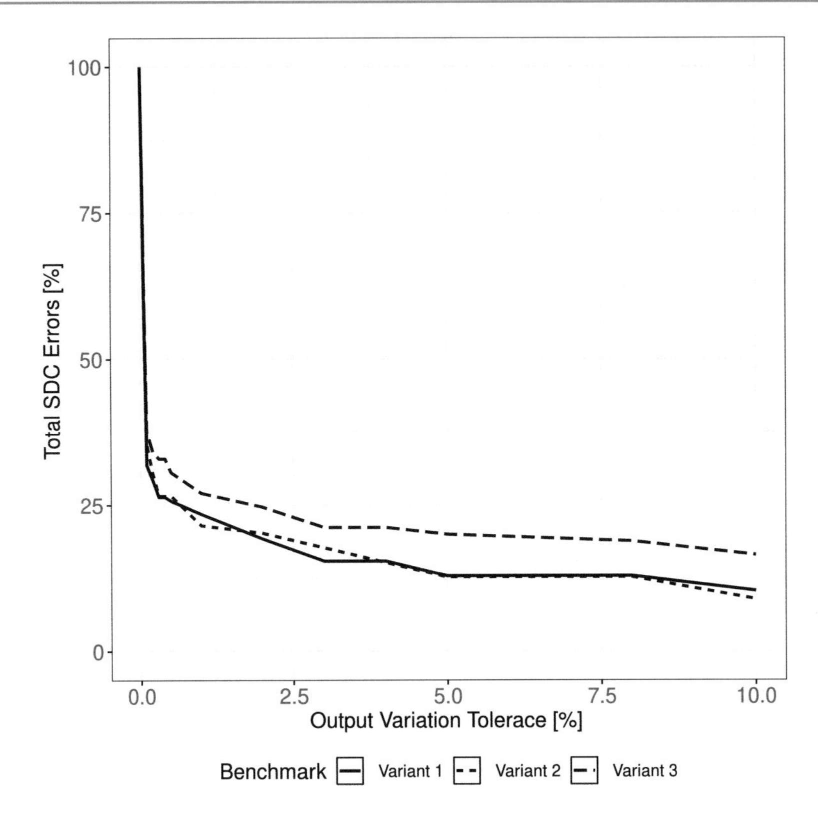

Fig. 7.13 Error occurrence drop in relation to output variation tolerance for Newton-Raphson benchmark. *Source* Author

The result analysis consists of a comparison between a golden execution of the application (i.e., with no fault injection) and the executions under fault injection. Three successive approximations and three ordinary computing algorithms are presented. Each one of those algorithms will be presented in three different versions. The first one is a bare metal application implementation, that is, executed on top of no OS. The second version runs an application with FreeRTOS operating system. The third version runs on top of Linux OS. The algorithms used as benchmarks are

- **Successive Approximation Algorithms**: They contain the benchmarks for the Newton-Raphson and Trapezoid methods, both used to approximate the result of the integral of a function, and QSolver, which presents the root computation of quadratic equations.
- **General Purpose Algorithms**: The Matrix Multiplication, Vector Sum, and Hanoi benchmarks represent ordinary algorithms. Those are normal matrix multiplication, vector sum, and the tower of the Hanoi puzzle solver.

Table 7.3 Error distribution for simulated fault injections using OVPSim

Execution		Errors [%]				
		General			Exceptions	
OS	Application	UNACE	SDC	HANG	Seg. fault	Unidentified
Bare Metal	QSolver	82.1	0.9	17.0	–	–
	Newton-Raphson	77.1	9.6	13.3	–	–
	Trapezoid	87.2	3.4	9.4	–	–
	Matrix Multiplication	70.1	19.5	10.4	–	–
	Vector Sum	65.2	23.6	11.2	–	–
	Hanoi	77.8	13.0	9.2	–	–
FreeRTOS	QSolver	43.0	9.6	47.1	–	0.4
	Newton-Raphson	74.5	8.7	15.9	–	0.9
	Trapezoid	58.3	10.4	31.2	–	0.1
	Matrix Multiplication	42.1	16.6	40.9	–	0.4
	Vector Sum	42.3	14.6	43.0	–	0.1
	Hanoi	24.5	40.8	30.7	–	4.0
Linux	QSolver	53.5	19.4	9.3	6.7	11.1
	Newton-Raphson	53.5	18.8	9.4	6.0	12.2
	Trapezoid	55.7	28.5	0.3	15.3	0.2
	Matrix Multiplication	45.4	37.0	4.8	12.3	0.5
	Vector Sum	43.6	40.2	4.9	10.9	0.5
	Hanoi	58.1	17.5	7.8	8.1	8.5

Table 7.3 presents the results for every type of error for each application and execution. Note that Exception errors are divided between segmentation faults and every other type (unidentified). This categorization is made because the majority of exceptions are usually segmentation faults, therefore it is interesting data.

Analyzing the bare metal results, it is clear that the successive approximation algorithms are much less susceptible to SDC errors than the three other applications. Comparing the worst-case scenario for the successive approximation algorithms (Newton-Raphson) with the best-case scenario of the other three (Hanoi), we have that the former is about 26% less susceptible to SDC errors than the latter. With the best-case scenario for the successive approximation (QSolver) compared with the worst case from the other three (Vector Sum), we find that the former may be up to 96% less susceptible to SDC errors than the latter. That means successive approximation algorithms may be from 26% up to 96% more reliable from SDC errors than ordinary calculation algorithms when executing bare metal. On the other hand, those algorithms are more susceptible to HANG errors, according to Table 7.3. Executing bare metal applications has a higher percentage of unace, which means they generated much fewer errors than Linux or FreeRTOS.

On the FreeRTOS cases, comparing the worst-case scenario for the successive approximation algorithms (Trapezoid) with the best-case scenario of the other three (Vector Sum), we have that the former are about 28% less susceptible to SDC errors than the latter. With the best-case scenario for the successive approximation (Newton-Raphson) compared with the worst case from the other three (Hanoi), we find that the former may be up to 78% less susceptible to SDC errors than the latter. With that data, it is observable that, on those tests, successive approximation algorithms are from 28% to 78% less susceptible to SDC errors than ordinary calculation algorithms when executing on top of FreeRTOS. FreeRTOS applications are much more susceptible to hangs than their Linux and bare metal counterparts. An application's distribution of errors differs when executing bare metal or on top of an operating system.

When executing on Linux, successive approximation algorithms did not have better fault tolerance than the ordinary computing applications but maintained a better susceptibility to SDC errors on average. It is clear, as already seen in Chap. 5, that the usage of an operating system drastically changes the fault tolerance. That is because the OS itself is a target for faults that may cause errors. In those simulations, one fault is injected per execution. Therefore, a big application will have a higher probability of having a fault injected during its execution, while for a small application, the chances are that the fault will be injected during the execution of some OS function. This is, in fact, in pace with the real-case scenario, where a short execution time means a lesser probability of having errors, as shown by [4]. The usage of successive approximation algorithms has its natural fault tolerance masked because of the OS criticality. Linux executions have to deal with segmentation fault errors, which are not present on FreeRTOS and bare metal. Nevertheless, those exceptions represent errors that were caught by the operating system. In the case of the absence of an operating system, those errors could manifest themselves as other types of errors.

7.3 Hardware ATMR

The hardware ATMR presented in this work uses data precision approximation (Sect. 6.2.1) and is implemented in the FPGA of the programmable logic layer of the Zynq-7000 APSoC. The methodology used to analyze this proposal is the onboard fault injection emulation, presented in Sect. 4.3. Physical experimental tests, such as radiation and laser fault injection, would also be proper evaluation methods. The DUT for the fault injection campaigns is the matrix multiplication of size 2×2. The ATMR designs presented in Sect. 6.2.1 are applied to the matrix multiplication designs. The benchmarks are assessed regarding their area consumption and inaccuracy introduced by approximate computing. Their implications are also discussed.

As presented in Sect. 6.4.1, six ATMR designs are implemented, varying the data precision of the operations. For simplification, the designs are named following the data precision of each redundancy module to simplify the analysis and data presentation (e.g., the ATMR design called "32-24-16" is composed of a module with 32-bit, one with a 24-bit and another with 16-bit precision data and operations).

Typically, when checking for errors caused by a fault, the output of the system would be compared with the one from an execution with no faults (called golden execution), and if a difference is found between those two values, we say an SDC error occurred. However, as discussed before, regarding the checker of ATMR methods, the error analysis methodology cannot be the same when dealing with approximate computing. As shown in Fig. 6.3, the approximation by data size reduction implies an inevitable accuracy loss. Because of that, there will always be a difference between the output of the full-accuracy golden execution and the approximated versions. In this study case, this difference will always be of at least 0.000303% (the lowest inaccuracy presented in Fig. 6.3).

The SDC error occurrence analysis must, therefore, take into account some acceptable differences between the fault-injected system output and the golden output. This acceptable difference between the two values is henceforth called *acceptance threshold* (ε). Section 7.3.1 presents the results for the random accumulated fault injections for two different acceptance thresholds: $\varepsilon = 0.01$ and $\varepsilon = 1$. Section 7.3.2 presents the exhaustive fault injection results for a $\varepsilon = 0.01$, i.e., an SDC is considered only if the difference between the system output and the golden value is equal to or higher than 0.01. The value of the acceptance threshold impacts the number of SDCs found by the analysis.

7.3.1 Random Accumulated Fault Injection

Figure 7.14 presents the results for the randomly injected accumulated faults on all the ATMR configurations for an $\varepsilon = 0.01$. The graph presents in the $y-$axis the reliability of the system and, in the $x-$axis, the number of faults accumulated on that point. The reliability is defined as the inverse of the occurrence of errors at a given number of accumulated injected faults (e.g., if the reliability at the point is 0.9 it means that 10% of the observed errors occurred with that number of accumulated injected faults or less). As expected, the ATMR configuration with three redundancies with 16-bit data is the one more reliable. It is clear that its curve is well detached from the other ones. Another expected result is the lower reliability of the full precision ATMR configuration (32-32-32) due to its larger area. However, the 32-32-32 curve is very similar to the 32-24-16 curve.

Figure 7.15 presents the results for the randomly injected accumulated faults on all the ATMR configurations for an $\varepsilon = 1$. That is a very high acceptance threshold that would only be acceptable in real-case scenarios where accuracy is not a strong concern. The ATMR

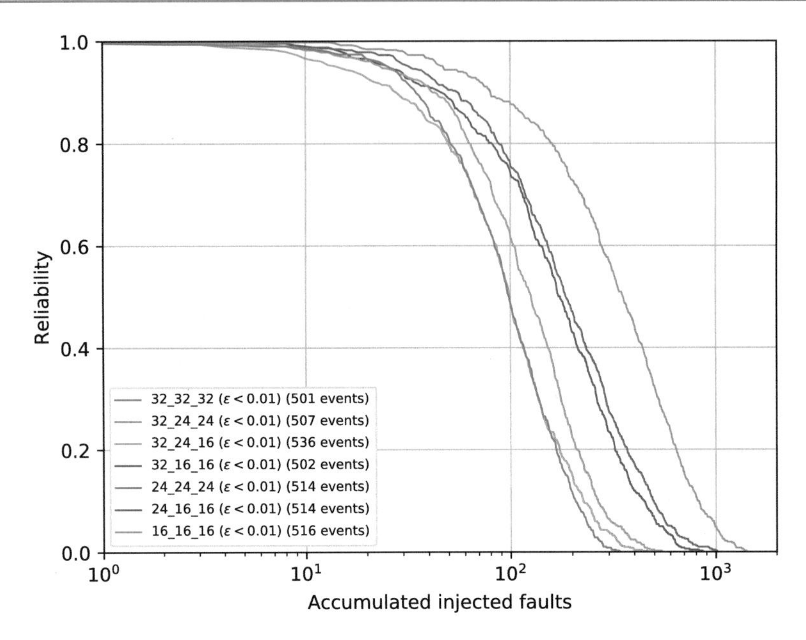

Fig. 7.14 Reliability for each ATMR configuration for an acceptance threshold of 0.01. *Source* Author

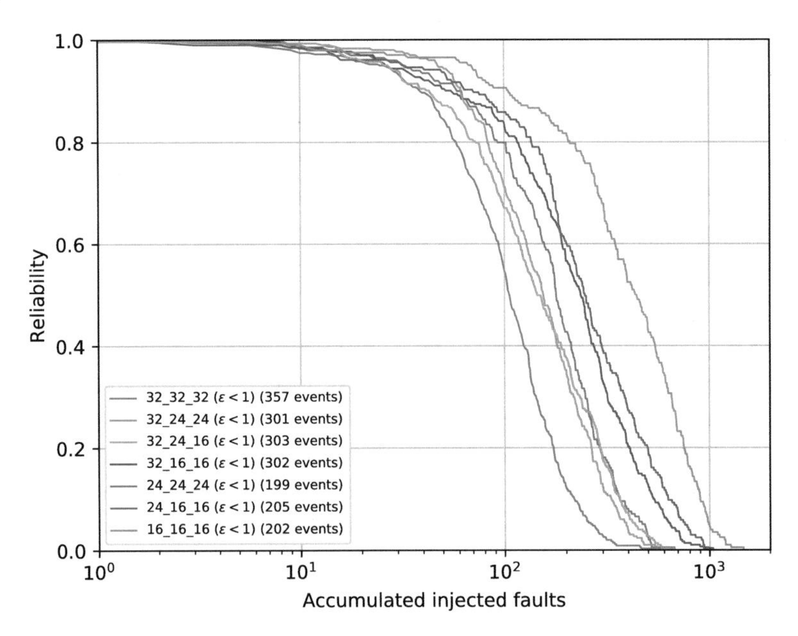

Fig. 7.15 Reliability for each ATMR configuration for an acceptance threshold of 1. *Source* Author

configuration with the highest reliability is again the one with three redundancies with 16-bit data. The difference between the two extremes of data precision configurations presented is evident. Nevertheless, the middle-term configurations seem to have similar reliabilities. It is also evident by comparing Figs. 7.14 and 7.15 that the behavior of the reliability curve is the same. However, the number of errors (number of events, on each figure legend) has dropped considerably.

The 32-32-32 ATMR configuration arises as the worst one in terms of reliability for $\varepsilon = 1$, distancing itself from the other curves. The fact that the 32-24-16 configuration is no more as bad as the 32-32-32 one indicates that this ATMR implementation is terrible when dealing with low-ε errors (Fig. 7.14), but is able to handle higher ones (Fig. 7.15). This is because this variable of the benchmark has to deal with the low precision of the 16-bit variables and the higher area of the 32- and 24-bit ones. Because the 32-24-16 configuration does not have two redundancies with the same precision, the 16-bit redundancy has a negative effect on accuracy without significant improvement in the fault tolerance with the area reduction.

7.3.2 Exhaustive Fault Injection

Table 7.4 presents the results from the exhaustive fault injections. Because of how the fault injection works, not all the injected faults affect the real DUT area. The fault injector affects a particular "rectangular" area of the FPGA layer, and because of the nature of the FPGA programming, not all of that area will contain the DUT. Therefore, the table presents the number of essential bits (which are the ones used by the design) and critical bits (the ones that caused errors when flipped) of the DUT. The last column of the table presents the variation of the number of critical bits in relation to the 32-32-32 TMR configuration.

As expected due to the previous observations, the 16-16-16 ATMR design is the one with the lowest number of critical bits. That is reflected in its high reliability concerning the other configurations. It is interesting to notice, however, that this ATMR configuration has a high percentage of critical bits in relation to essential bits. It indicates that a design of a smaller

Table 7.4 Exhaustive onboard fault injection emulation results for a 2×2 matrix multiplication

TMR design	Essential bits	Critical bits	Critical bit variation (†) (%)
32-32-32	540454	7126	0
32-24-24	355164	3296	−53.47
32-24-16	299456	4016	−43.64
32-16-16	228122	4178	−41.36
24-24-24	305093	6343	−10.98
24-16-16	165172	3724	−47.74
16-16-16	88253	1764	−75.24

(†) In relation to the 32-32-32 TMR design

area tends to be more reliable, even if a higher percentage of this design is critical. This idea is also backed by the fact that the 32-32-32 and 32-24-16 ATMR configurations are the ones with the worst reliability (as presented in Sect. 7.3.1) and also a high number of critical bits.

The 24-24-24 ATMR configuration is the one with the second highest number of critical bits (32-32-32 being the one with the highest). Given this fact, it could be expected that it would also be the one with the second worst reliability. That, however, is not the case. Both Figs. 7.14 and 7.15 show that the 24-24-24 configuration is actually between the worse and the best ones, which proves that the precision and accuracy of the design also play a significant role in the system reliability.

7.4 Software ATMR

As detailed in Sect. 6.4.2, successive approximation algorithms always produce an approximated output value. As discussed before, on approximate computing, a small deviation from the golden value is reasonable and thus accepted. Therefore, similar to the presentation of the results from Sect. 7.3, the number of errors and masked errors will always be presented concerning a given acceptable threshold of difference between the ATMR task values (and the golden value). For example, a threshold of 2% means that the error is less than 2% different from the golden value. This is not exactly the same threshold system presented in Sect. 7.3: this one is based on a percentage of difference, not on absolute values. Results are presented for three different thresholds. The ATMR configurations are the same ones already presented in Sect. 6.4.2 and are evaluated concerning their fault coverage and execution overhead. However, they are now tested by combining two approximation methods: loop-perforation and data precision reduction. Each of the ATMR configurations presents two versions, one implemented with single-precision and the other using double-precision variables. This ATMR technique was tested under laser fault injection on the L1 data cache memory, following the methodology described in Sect. 4.6.

Figures 7.16, 7.17, and 7.18 present the “Error Distribution” of the ATMR tasks (i.e., the number of ATMR tasks with errors) applied to the single-precision version of the Newton-Raphson algorithm. They respectively present data for ≈0%, 2%, and 5% difference thresholds between the outputs of the tasks and the golden value. The ≈0% data presented actually stands for a difference of 0.000013%, which is the difference between the values from the 71- and the 14-iteration executions (without errors). It is written as ≈0% for simplification, and because it is the maximum difference that will always be present due to the usage of approximation in this application. Data is presented in percentage and calculated concerning the number of the benchmark executions that had any difference between the task output and the expected golden value. For example, in Fig. 7.16 the white bar on the graphs (called “1 of 3”) presents the percentage of the executions with errors that contained an error in one of the three ATMR tasks, considering a ≈0% difference threshold between the outputs of

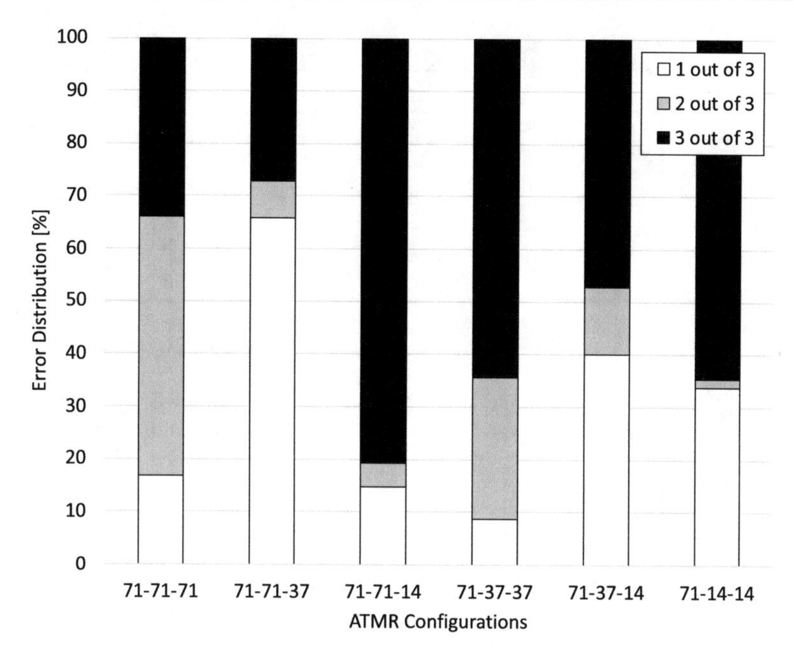

Fig. 7.16 Number of ATMR tasks with errors for a ≈0% difference threshold between the task outputs and golden value, on the single-precision version of the Newton-Raphson algorithm. *Source* Author

the tasks and the golden value. To gather this data, the output of each task is compared to the golden value and checked for errors.

Figure 7.16 shows that a considerable amount of errors are not corrected by the ATMR (because most cases presented two or more tasks with errors). This result is expected because of the natural variation of approximate computing algorithms outputs. When using single-precision, the 71-71-14 ATMR is the one with the highest percentage of errors affecting three out of three tasks. Two factors can explain it. First, the 14-iteration task is the one most susceptible to faults. Secondly, the 71-iteration task is the one with the higher execution time. A higher execution time means more exposition to faults (because the laser pulse frequency is constant for all benchmarks). Those two factors contribute to a very inefficient ATMR configuration.

In Figs. 7.17 and 7.18, the "Vanished" bars represent the amount of errors that are no more present when the difference threshold increased (respectively from ≈0% to 2% and from ≈ 0% to 5%). Figure 7.17 shows that increasing the acceptable difference threshold between the outputs and the golden value not only masks some errors but also decreases the number of erroneous tasks. The same behavior is also observed in Fig. 7.18, where the difference threshold increased to 5%. Comparing the data from Figs. 7.17 and 7.18, it becomes evident that the amount of vanished errors cease to increase at a certain point. It indicates that there

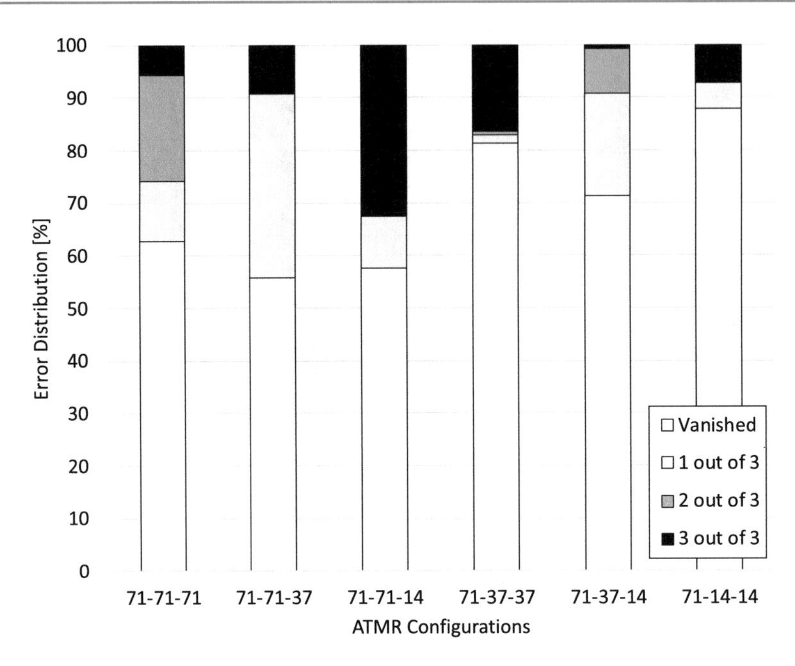

Fig. 7.17 Number of ATMR tasks with errors for a 2% difference threshold between the task outputs and golden value, on the single-precision version of the Newton-Raphson algorithm. *Source* Author

may be an optimal difference threshold point, capable of providing good fault tolerance while not compromising too much the output accuracy.

Figures 7.19, 7.20, and 7.21 present the error distribution of the ATMR tasks applied to the double-precision version of the Newton-Raphson algorithm, respectively, presenting data for ≈0%, 2%, and 5% difference thresholds between the outputs of the tasks and the golden value. Once again, the "Vanished" bars in Figs. 7.20 and 7.21 present the amount of errors that are no more present when the difference threshold increased. Comparing Figs. 7.19 and 7.16, using double-precision variables makes a ≈0% difference threshold ATMR even less appropriate. The number of executions with errors affecting two and three tasks is more relevant in that case. However, increasing the difference threshold between the outputs and the golden value highly increases the fault masking capability of the technique. Figure 7.20 shows that a 2% threshold is enough to provide a good fault masking. Figure 7.21 shows that increasing the threshold to 5% does not improve the fault masking performance very much in comparison with a 2% threshold.

Table 7.5 presents the percentage of masked errors for three thresholds of difference between the ATMR voted values and the golden value. Different from the data shown in Figs. 7.16, 7.17, 7.18, 7.19, 7.20, and 7.21, this now concerns the value voted by the ATMR, not the outputs from the tasks. As discussed before, the more the iterations successive approximation algorithms have, the more the fault-tolerant we expect them to be. However, some

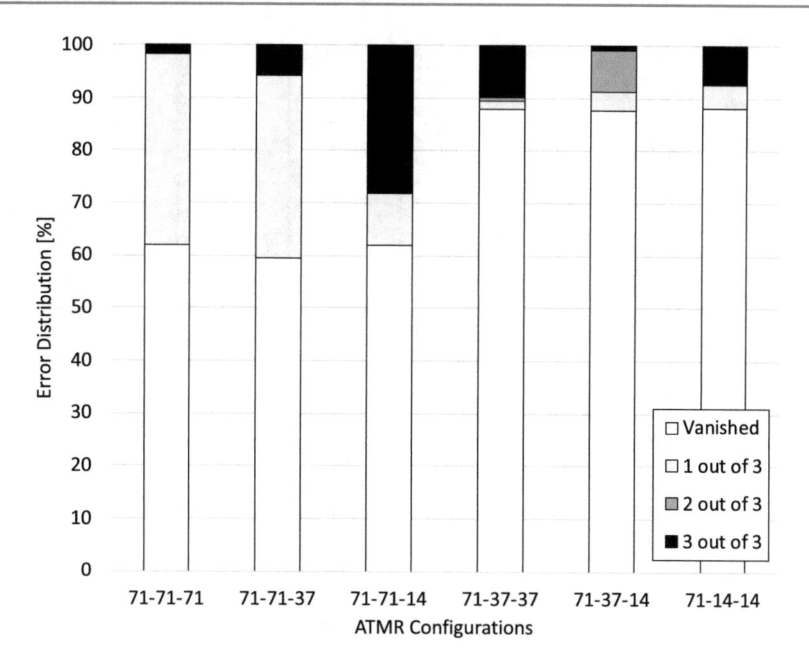

Fig. 7.18 Number of ATMR tasks with errors for a 5% difference threshold between the tasks outputs and golden value, on the single-precision version of the Newton-Raphson algorithm. *Source* Author

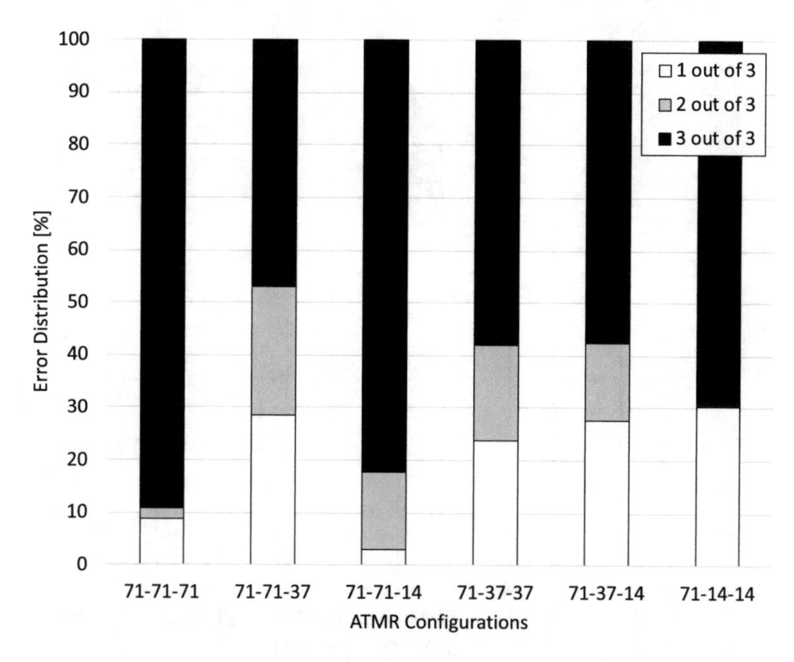

Fig. 7.19 Number of ATMR tasks with errors for a $\approx 0\%$ difference threshold between the task outputs and golden value, on the double-precision version of the Newton-Raphson algorithm. *Source* Author

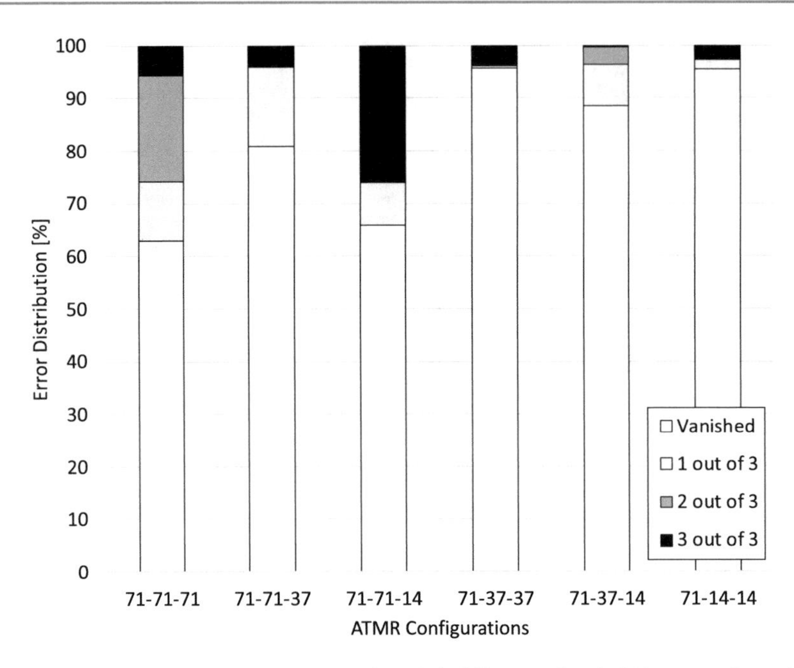

Fig. 7.20 Number of ATMR tasks with errors for a 2% difference threshold between the task outputs and golden value, on the double-precision version of the Newton-Raphson algorithm. *Source* Author

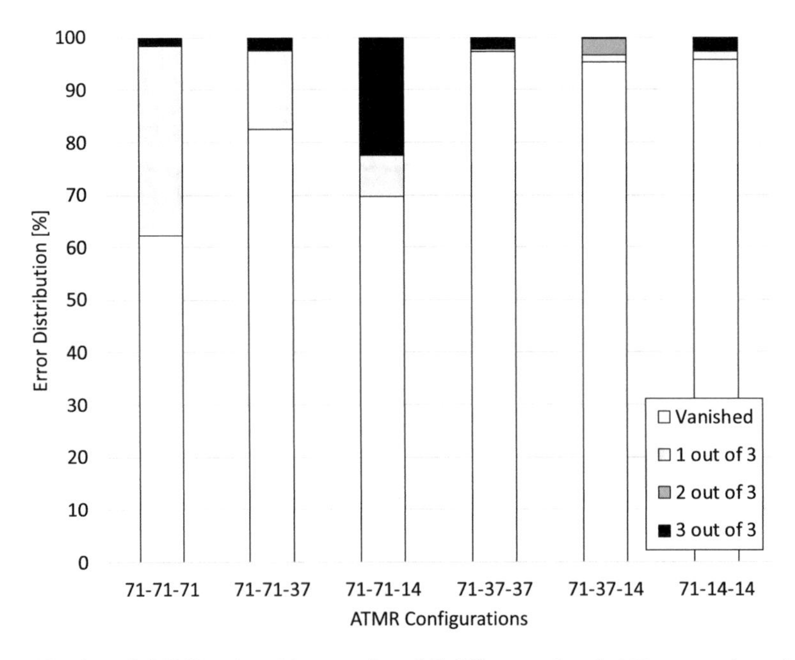

Fig. 7.21 Number of ATMR tasks with errors for a 5% difference threshold between the task outputs and golden value, on the double-precision version of the Newton-Raphson algorithm. *Source* Author

Table 7.5 Error masking for each ATMR configuration variating thresholds

ATMR config.	Single-precision			Double-precision		
	≈0% Thres.	2% Thres.	5% Thres.	≈0% Thres.	2% Thres.	5% Thres.
71-71-71	17.37	76.02	97.90	9.30	87.30	87.35
71-71-37	66.80	91.06	94.72	34.68	88.36	88.36
71-71-14	14.84	66.06	66.06	4.29	98.73	98.73
71-37-37	8.82	82.56	89.19	30.09	92.19	94.62
71-37-14	40.03	90.83	91.31	27.67	80.45	80.45
71-14-14	33.82	94.97	94.97	31.88	99.96	99.98

unexpected results are present. Such is the case of the 71-14-14 ATMR configuration, due to its high performance both for the single- and double-precision implementations. Because more iterations usually mean more fault tolerance, this is non-intuitive. Nevertheless, it can be explained by the execution time of this benchmark. It is the one with the lowest overhead (Table 6.4), being subject to fewer fault injections than the others. Literature shows that a high execution time implies low fault tolerance, once the system is exposed to more faults, particularly in radioactive environments [3, 4].

Table 7.5 shows that by increasing the threshold, the ATMR was capable of masking many more faults. Even a small difference threshold of 2% is enough to make some configurations mask more than 90% of the errors. The ATMR configuration capable of masking most errors with single-precision with a high threshold is 71-71-71. However, this configuration performs very poorly for a small threshold. This is probably due to the fact that this is the configuration with the highest execution time and therefore is subject to more faults per execution. In this case, increasing the number of iterations would, instead of improving the fault tolerance (by making the output converge), make it worse (because of the high execution time). The 71-14-14 configuration is the best one at double-precision, and it reaches a good error masking even for a 2% difference threshold. The double-precision implementations have worse performance than the single-precision ones for the $\approx 0\%$ acceptable difference threshold. Nevertheless, increasing the threshold increases the error masking faster than it did in the single-precision cases.

7.5 PAED

The PAED technique proposed in Sect. 6.5 is evaluated under laser fault injections, following the same laser configuration detailed in the other experiments of this work. The experiments consisted of injecting faults both at the OCM and L1 data cache memory of the DUT (which

Table 7.6 Details of the benchmarks used to evaluate PAED

Application		Approximation method	Exec time [c.c]	Processed data [kB]	Approx. checker threshold
Cubic	Std.	–	472850	152.625	–
	Apx.	Data precision	439646	76.312	0.00001
Matrix Multiplication	Std.	–	1253588	56.250	–
	Apx.	Data precision	1206104	28.125	0.0006498
FFT	Std.	–	1629622	16.000	–
	Apx.	Data precision	1650040	8.000	0.0065
Trapezoid	Std.	–	4521028	1.028	–
	Apx.	Loop-perforation	3519542	1.028	0.0248
Newton-Raphson	Std.	–	4145720	0.250	–
	Apx.	Loop-perforation	4123290	0.250	0.0248

is the already detailed dual-core ARM Cortex-A9 processor embedded at the Zynq-7000 SoC).

The benchmarks used in the experiments to evaluate this proposal are the Trapezoid and Newton-Raphson numeric methods that were already presented and discussed in Sect. 7.2.2, and two new ones: cubic and fast Fourier transform (FFT). The cubic benchmark comes from the automotive package of MiBench [1]. It consists of calculations of cubic equation solutions, integer square roots, and angle conversions. The code makes use of many trigonometrical functions such as sin and cosine calculations, which are by themselves already approximate. The FFT also comes from the MiBench benchmark suite, and consists of a traditional calculation of an FFT, taking as input vectors representing real and imaginary values. The outputs are stored in the same manner.

Table 7.6 presents the details of the benchmark applications used in this work. The data presented in the table concerns a single execution of the application as a task of the system, thus the low execution time. As discussed in Chaps. 2 and 6, there will always exist a small difference between the results of the approximate redundancies and their non-approximate counterparts. The last column of Table 7.6 presents the threshold of the approximate checker implemented at the PAED technique in absolute values. This threshold is intended to accept the highest difference between $task_0$ and $task_1$ (remember Fig. 6.6) outputs, caused by the usage of approximation, in the absence of faults. The two approximate computing methods used are loop-perforation and data precision reduction. In the data precision reduction, the standard version of the algorithms uses 64-bit floating-point data, while the approximate version uses 32-bit data.

Results for the approximate error detection technique are presented concerning the error detection rate of the technique applied to the benchmarks evaluated. For each benchmark, the technique always executes the standard version at CPU0. The redundancy executed at the CPU1 is either the standard (traditional DWC) or the approximated one (PAED). The y-axis of the figures presented in the following subsections indicates the version of the algorithm being executed by CPU1. Data is presented and analyzed categorizing the experiment results into three types:

- **Detected Errors**: It is calculated comparing the total number of erroneous CPU0 calculation outputs (i.e., the value is different from the golden value) and how many of those were found by the error detection technique;
- **False Detection**: It presents the percentage of errors detected by the technique that are actually not errors (i.e., the technique issues a warning to the system, but the output from the $task_0$ at CPU0 is actually correct);
- **Undetected Errors**: When there is an error at the $task_0$ output from CPU0, but the technique did not issue a warning to the system, we say we have an undetected error.

Those categories are presented in percentage of total occurrences, the total occurrences count being the sum of events of the three types.

Figure 7.22 presents the results for the error detection rates obtained by the proposed technique of laser fault injections at the OCM. For cubic, trapezoid, and Newton-Raphson benchmarks, the usage of approximate redundancies had little impact on the error detection when compared to the non-approximate redundancy. In those cases, both PAED and the traditional DWC detected almost all errors. This, however, is not the case with matrix multiplication. The results from the matrix multiplication present poor performance for the standard redundancy. The proposed PAED technique, however, showed a much better error detection than the traditional DWC, detecting almost all errors. The same behavior can be observed when comparing the results from the FFT benchmark, on which the usage of approximate redundancy and the PAED technique highly increased the error detection.

Figure 7.23 presents the error detection rates concerning the laser fault injections at the L1 data cache. In this scenario, the traditional DWC error detection performed poorly for the cubic benchmark. The PAED technique presented a much better error detection but still lower than the ones attained in the other applications. This may be explained by an essential characteristic of the cubic benchmark: it implements approximation by data precision reduction. Because the L1 data cache memory is smaller than the OCM, data size reduction impacts the reliability of the system much less. Matrix multiplication also has data precision reduction as its approximation method. However, it is much more memory-intensive than the cubic benchmark, for it has to constantly read from matrix inputs and write into the output matrix. This line of deduction can also explain why the usage of approximation increased the error detection at the matrix multiplication so much, repeating the observations from Fig. 7.22. The L1 data cache memory is a critical area of the system. That is why

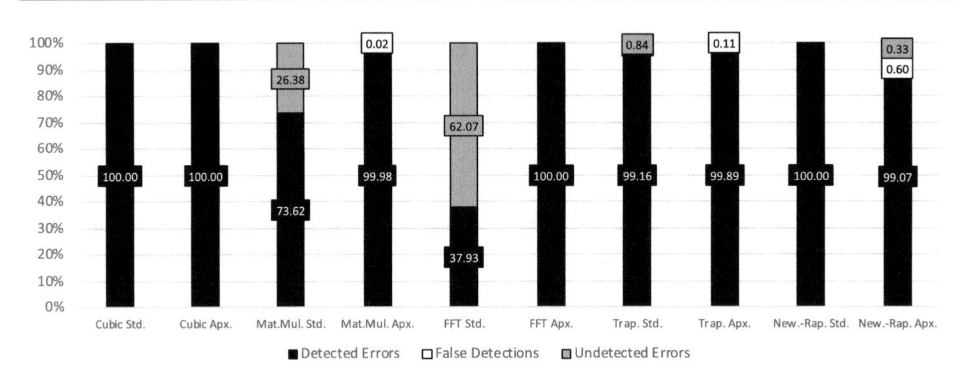

Fig. 7.22 Error detection rates for laser fault injection at the OCM memory. *Source* Author

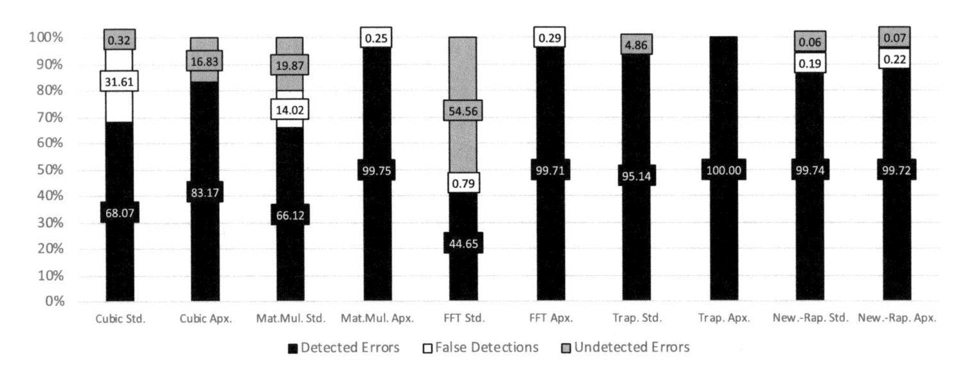

Fig. 7.23 Error detection rates for laser fault injection at the L1 data cache memory. *Source* Author

data precision reduction presents such an important improvement in the reliability of the benchmarks under fault injections in this ROI. This is also proved by the results from the FFT benchmark, where the approximation once again caused a high increase in the error detection rate, and the standard DWC presented terrible results.

It is also important to remember that the OCM memory holds the stack and the heap of the applications, while the L1 data cache holds the more frequently used data. This explains the differences between the results from Figs. 7.22 and 7.23. Indeed, the results from 7.23 presented overall a lower error detection. Memory-intensive algorithms such as matrix multiplication are much more prone to errors caused by faults affecting the memory than less intensive ones such as trapezoid. That explains the significant differences between the approximate and standard versions of matrix multiplication and FFT under both laser fault injections. Even though the cubic benchmark uses more memory (Table 7.6), matrix multiplication has more memory accesses and interdependencies, due to the nature of the operation.

References

1. M.R. Guthaus, J.S. Ringenberg, D. Ernst, T.M. Austin, T. Mudge, R.B. Brown, MiBench: a free, commercially representative embedded benchmark suite, in *Proceedings of the Fourth Annual IEEE International Workshop on Workload Characterization. WWC-4 (Cat. No.01EX538)* (2001), pp. 3–14
2. R. Leveugle, A. Calvez, P. Maistri, P. Vanhauwaert, Statistical fault injection: quantified error and confidence, in *2009 Design, Automation Test in Europe Conference Exhibition* (2009), pp. 502–506
3. H. Quinn, Challenges in testing complex systems. IEEE Trans. Nucl. Sci. **61**(2), 766–786 (2014)
4. G.A. Reis, J. Chang, N. Vachharajani, S.S. Mukherjee, R. Rangan, D.I. August, Design and evaluation of hybrid fault-detection systems, in *32nd International Symposium on Computer Architecture (ISCA'05)* (2005), pp. 148–159
5. J. Tonfat, L. Tambara, A. Santos, F. Kastensmidt, Method to analyze the susceptibility of HLS designs in SRAM-based FPGAs under soft errors, in *Applied Reconfigurable Computing*, ed. by V. Bonato, C. Bouganis, M. Gorgon (Springer International Publishing, Cham, 2016), pp. 132–143

8 Final Conclusions and Remarks

8.1 Summary

In this book, we tried to summarize the state of the art in approximate computing and its relation to safety-critical algorithms and fault tolerance. We also presented various methodologies and metrics most commonly used to assess them. Finally, the last part of the book was dedicated to putting into practice the theory studied previously and putting all the presumptions under test. The results show to be promising for approximate computing and its interaction with safety-critical systems that work under hazardous circumstances such as those from space and other radioactive environments.

8.2 Approximate Computing

The idea of saving resources by reducing the amount of effort spent on a task is not a new one. Humanity itself has been doing it for times unimaginable. So, applying that concept to computer science is only natural. And from all that we have seen in this book and the literature supporting it, it works. From approximation strategies to naturally approximate numerical algorithms and approximate fault tolerance techniques, approximating computing proves to be a good answer for resource utilization savings.

Even though we discussed in Chap. 2 a plethora of approximation techniques, the concept of approximate computing can be expanded way beyond that. The Taylor Series approximation, for example, defined and studied in Chap. 6, is not exactly a method for approximating algorithms. Instead, it is by itself a means of achieving the desired output. The same is valid for the Newton-Raphson algorithm, for example, also detailed in Chap. 6 and extensively assessed and discussed in this book. Those numerical methods are examples of a whole class

G. S. Rodrigues et al., *Approximate Computing and its Impact on Accuracy, Reliability and Fault-Tolerance*, Synthesis Lectures on Engineering, Science, and Technology,
https://doi.org/10.1007/978-3-031-15717-2_8

of algorithms that are by nature approximate and are often applied to safety-critical systems as part of their functionality.

The question of whether any function can be approximated with the use of numerical methods or natural approximation is out of the scope of this book. However, some of our results indicate that this type of algorithm does indeed possess a reliant nature and an inner fault tolerance in the form of auto-correction. Therefore, an intelligent designer might wonder if the solution he is developing might be better attained with the use of numerical approximation.

8.3 Safety-Critical Systems, Reliability, and Approximate Fault Tolerance

The safety-critical systems industry is a conservative one, as it should. This type of system frequently deals with human lives and the safety of others. They are also often costly: a catastrophic error in a satellite, for example, means the loss of millions of dollars in investment. So, all things considered, it is only natural for a designer of this type of system to "play safe" and rely on time-proven strategies.

Nevertheless, Chap. 6 showed that approximate computing algorithms are not only as reliable as any other ordinary but actually even safer. The results presented in Chap. 7 also supported this claim. Moreover, they show that approximate computing, when applied to fault tolerance techniques, might make them more reliable given the right circumstances.

Approximating a fault tolerance technique can also be more straightforward than one might expect. In this book, we discussed a plethora of approximation methods applied to hardware and software. On both stacks, there are approximating methods already extensively discussed in the literature that make the job of approximating an algorithm much easier. As presented in Chap. 2, some of those strategies consist of merely importing a library that helps the designer easily define the size of its floating-point data in an HLS or VHDL project, for example. In software, implementing an approximation by data precision reduction and loop-perforation can also be extremely simple, although this last one might have its impact limited by the nature of some codes. If successfully implemented, results presented in Chap. 7 and in an extensive amount of publications indicate that they are very successful in providing fault detection and error masking by a fraction of their non-approximate counterparts.

8.4 Experimental Results

Table 8.1 presents a summary of the results obtained in this book's experiments. Note that it is just a summary and does not specify the different methodologies used for getting the results. This table can be misleading by itself because it only shows the best-case scenarios and because the different experiment methodologies analyze fault tolerance in different

Table 8.1 Summary of results

Techniques		Cost reduction			Impact on quality and accuracy	Fault tolerance	
		HW area	Exec. time or latency	Memory footprint		Error detection	Error masking
Approximation	Data precision reduction	–	–	−50%	Loss of 8 decimal digits	–	–
	Successive Approximation	Up to −75% DSP usage	Up to −99.4%	–	User-defined (‡)	–	Up to 87% SDC reduction
	Taylor series	Up to −95% DSP −97% FF −97% LUT	Up to −95%	−50% (†)	Up to 71% degradation (in software, very few terms)	–	Up to 97% essential bits reduction
Fault tolerance	Hardware ATMR	Up to −75% DSP −70% FF −91% LUT	Up to −40%	−50% (†)	User-defined	–	Up to 75% critical bits reduction
	Software ATMR	–	Up to −58%	−50% (†)	User-defined	–	From 66.8% (≈0% threshold) to 99.9% (5% threshold)
	PAED	–	Up to ≈ −30%	−50% (†)	Worst case: second decimal Best case: fifth decimal (‡)	Up to 100% detection	–

(†) When used with data precision reduction
(‡) Highly dependent on the implementation and system requirements

ways. It shall be read taking into account all the discussions and details from Chap. 7. For the ATMR on hardware, for instance, the table only presents the results regarding the exhaustive fault injection because the ones from the randomly accumulated fault injections are too complex to put on a table. One of the points to take into account is, for example, the execution time reduction from PAED. It could be much improved with the evaluation of different approximation intensities. Taylor series error masking only presents data regarding essential bits because no fault injection was performed on it. Some data presented in Chap. 7 had to be left out of the table due to complexity, such as the results from the fault injection on different operating systems and its comparisons.

Successive approximation arises as a promising approach to approximate computing. The comparison between the Trapezoid and Simpson benchmarks shows that the number of iterations alone is not enough to assure a method will achieve good resilience. The algorithm itself has a significant impact on fault tolerance. Therefore, studying this kind of approximate computing algorithm is essential before it can be applied to safety-critical systems as reliable software. However, the number of iterations does affect the fault tolerance. Even for a benchmark with a relatively low number of iterations, such as Newton-Raphson, its impact on fault tolerance is noticeable.

Results also show that the Taylor series approximation is capable of achieving excellent accuracy with a small number of Taylor series sum terms. The performance and cost of the Taylor series approximation depend on the target algorithm accuracy constraints. It proved to be able to provide fast and low-cost approximations for systems with low limitations as well as good approximations (up to 100% accuracy) for those who need it (and can pay for its cost). Hardware implementations appear to be either slower than embedded software or consume too many resources.

The proposed ATMR by successive approximation approach decreases the execution time overhead compared to the classical TMR while keeping an acceptable fault masking rate. Using single- and double-precision floating-point variable types had an impact on the error masking of the method, but the general behavior remained the same. All the benchmarks showed a trend of having a significant drop in the number of SDC errors for small output variation tolerances. It shows that most of the SDC type errors affecting successive approximation algorithms are not significant. Many applications that use this kind of algorithm may tolerate minor variations in the output without a problem. For those applications, successive approximation arises as the perfect method for approximate computing.

The ATMR by data precision reduction proved capable of generating implementations with lower area usage while maintaining good accuracy. In the worst-case study scenario, the accuracy remained higher than 99.96%. The area reduction provided by this method could be used to provide better performance. A multitude of design strategies could be applied to use the now vacant FPGA area and deliver more parallelism. It is also worth noticing that the proposed ATMR method provided lower latency than the more accurate design option. The fault injection experiments prove that approximating the system impacts its reliability. The smaller, approximated circuits presented higher reliability than the bigger, more accurate ones. The results also indicate that approximate computing improves the system reliability not only by making it smaller but also because of the nature of the approximation.

The results indicate that the proposed parallel approximate error detection (PAED) technique presents a relatively better improvement in error detection for memory-intensive tasks than on processor-intensive tasks. This, however, does not imply a limitation of applications for the proposed technique. The fact that the traditional DWC was capable of high error detection does not invalidate the PAED technique. On the contrary, PAED was capable of maintaining proper error detection (even further improving it) while presenting a lower implementation cost on both execution time and memory footprint. The technique is therefore attractive for systems that deal with data freshness, such as real-time systems. From a real-case scenario point of view, this type of error detection would be great to alert an aircraft pilot that a particular value is unreliable. Because of the data freshness time window, the pilot would take the best attitude to deal with this problem in the safest manner until a new refreshed data is generated, this time with a reliable value.

It is nevertheless essential to notice that not all systems may be able to make use of approximate computing. Some applications cannot afford any inaccuracy, which makes those applications out of the scope of the proposed approach. Further theoretical analysis proved the inherent capability of successive approximation to handle faults before they become errors.

Printed in the United States
by Baker & Taylor Publisher Services